A Short Presentation of the History of Science

A Short Presentation of the History of Science

First Edition

Llewellyn Pearce

Copyright © 2017 by Llewellyn Pearce.

ISBN:	Softcover	978-1-5434-4030-0
	eBook	978-1-5434-4031-7

All rights reserved. No part of this book may be reproduced or transmitted in any form or by any means, electronic or mechanical, including photocopying, recording, or by any information storage and retrieval system, without permission in writing from the copyright owner.

Any people depicted in stock imagery provided by Thinkstock are models, and such images are being used for illustrative purposes only.
Certain stock imagery © Thinkstock.

Print information available on the last page.

Rev. date: 08/01/2017

To order additional copies of this book, contact:
Xlibris
1-888-795-4274
www.Xlibris.com
Orders@Xlibris.com
751469

CONTENTS

Preface vii
Introduction ix

History Of Science 1
Introduction To Science For Teaching Young Children 2
Ancient Science And Mathematics 19
The Age Of Enlightenment 32
Space Missions 49
Modern Developments In The Late 20th-Century And The Post-2000 Year Time Period 57
History Of Science For The Future 66
History Of Science And Medical Discoveries 74

PREFACE

This First Edition presents a quick look at the history of science, and will be followed by a second edition with greatly increased coverage of science covering Galileo, Newton, and later development in science, with attention to mathematical formulas, largely avoided in many science periodicals, and even in the World Book Encyclopedia.

INTRODUCTION

The United States of America led the world in science after WW-II, but should not lose that leadership. The local public television and FM radio networks should begin to carry more science programs. I encourage parents to read to their children about science, starting at an early age. To help in the reading to children, public television programs could be of immense importance. The public TV and FM networks should open special channels for science. Good science programs could be instrumental in reversing the present trends in (non-science) programming, at least on such new science channels. The high cost of university education may be reduced through the use of the internet. Internet usage for academic study should be started in, or even before, grade school, with emphasis on mathematics and modern science training. My plan for the future is to develop a continuing series of presentations of science history and discussions aimed towards teaching children, teenagers, and adult parents and friends, for both girls and boys. Develop and save science study DVD's by you, your parents, and your best teachers, and University professors.

New science public TV channels would help to introduce science and mathematics, starting the day with children's science programs, and later for adults. Development of young children needs to be started earlier than possible with the usual school year, using the internet with specially designed internet pre-school programs. Programs should go on all year, and should include library DVD disc presentations. DVD programs could be customized and adapted to the interests of the individual, child and adults.

There are several main objectives of this book. The first objective is to present a detailed discussion of science developments over the past 4000 years. The second objective is to provide a well-documented outline of science advancements for use in evening and summer reading in family-oriented study sessions for presenting science to young children to increase science interest and understanding by our youth, via family discussion groups, libraries and museum visits. Science developments have and always will be vital for the world economy and the availability of good jobs. The roles provided by public radio and television should include thorough science presentation programs, not merely the present excessive and repetitious entertainment set of programs. The public should encourage more and better science presentations.

The role of public television should include good science programs rather than so many violent and purely entertainment programs. The serious problem with so many violent programs on the media is that it teaches violent behavior to our youth, and violent programs become more exciting than programs stressing normal behavior patterns, and science.

One impact of a lack of interesting programs on science is that science budgets can be severely cut back during difficult economic times. Lack of science interest by the public can lead to reduced support by political leaders for science developments. Political leaders should never refer to current science as anti-business. New scientific discoveries usually lead to increased business, and therefore should be presented to the general public, as well as to young children, both girls and boys.

For several recent decades, politicians, business leaders, and lawyers seem to be lacking in scientific training. For example a new leader in the University of California is a brilliant attorney, not trained in advanced science studies. That person is capable of fast learning, and it is not too late for that leader and others to start taking on a personal program for science study. The problem may be the result of less interest in academic science, which should include medicine, nursing care, mathematics, Earth sciences, chemistry, and physics. Good science study should begin with history, especially science history.

This humble author encourages you as an intelligent family to build a small home library with important science books and saved articles and your personally written essays on science with help from the internet and library.

Some libraries limit the acquisition of academic books, with the excuse that the public has limited interest in science. For good books covering academic subjects, families should visit libraries in city colleges. Some public libraries avoid acquiring books on academic subjects. Families need to visit city college and university libraries. Browse used college book stores for older textbooks.

Books and articles on science in some libraries are becoming scarce, due to the cost of printing, storage, limited public library shelf space, and delivery cost. Many important books are becoming difficult to find, and may become out of print, and no longer available, except from a good used book store. Science materials are particularly hard to find. Some libraries limit purchasing good science books. Books required by professors for student purchase are rarely on small library book shelves. Such books may be found in school libraries, and in used book stores.

Develop good science materials on DVD's by our best scientists. Browse old book stores for bargain books. Classic editions and used textbooks are always a good investment. College textbooks should be saved; they are among the first to go out of print, and will become even more valuable later. A DVD science library should be developed and made available in public libraries.

Magazine publishers are considering releasing previous year's issues on DVD. The US Courts seem to approve that practice, as acceptable. Many scientific periodicals already are available on DVD's.

An example of important publications not being retained by many public libraries, include:

U.S. Nautical Almanac, published yearly in book form: much reduced in size after about 1990

The Astronomical Almanac

The Encyclopedia Britannica

Newsweek magazine—no longer available in printed form

Scientific periodicals – may become available only on DVD's, and no longer available in printed form Nature, and Science periodical magazines are considering change to DVD only.

Science magazines and other publications in the DVD format are presently being developed and may be found if you ask your library reference librarian for guidance.

The present cost of an important book from Amazon.com or Half.com has become reasonable. How long will such bargain prices be available is unknown. Used book prices may be lower than Amazon. Most librarians release good, but older books and magazines for free, or at low prices. Build your library early and gradually; it should become very valuable in the future. Make your library accessible to your children, grandchildren, and your many friends. Save a section of science books and science DVD's from local libraries in your home in a permanent collection to be handed down from generation to generation, and shared with friends.

This book presents a somewhat detailed history of science from the early Asiatic, Arab and Greek periods to modern times. This book is in the form of a detailed study guide for parents and young students. It is an attempt to encourage the early-age teaching of science to young girls and boys by parents. Science knowledge is important for all intelligent students. Schools usually defer science studies to high school. Family trips and family group discussion sessions on science is the main way to introduce science to young children. NOVA science programs are important, but are too few. I think that girls as well as boys need to read about science starting at the age of 6 or 7 or earlier by reading and parent-study groups, with lots of educational programs, and less entertainment. Leave entertainment to commercial television. Put science programs back on public television. Science study is now more important than ever, considering the high cost of science education and lack of classroom availability.

If your children are still very young, this is the time for you to start reading to them about science, for your family and friends are the best early teaching groups for your children's development. The United States of America and the world needs more scientists, engineers and medical workers. More and higher paid jobs are available for scientists, engineers, geologists, medical doctors and nurses. The cost of university education is rapidly becoming so expensive that only the very wealthy can afford it. Home reading sessions leading to future science study in a university obtained by a full scholarship, is just about the only way to remedy the high costs of advanced schooling.

Benjamin Franklin said that a penny saved is a penny earned (and available to earn interest, long term). I could re-state that proverb as 'A great used book obtained from a library book sale will appreciate even faster than a penny'.

After World War II, science education in Junior college to graduate school was financed by most governments. Starting in the 21-st century, funding for advanced education in the sciences is obtained more frequently by scholarships, or costly student loans. Government funding for college tuition is limited, except by scholarship. To get a good scholarship, science studies should begin as early as possible and such early study might well depend on family-oriented reading and group learning sessions. Many university courses may become unavailable due to filled enrollment, and may be very expensive. Some of such courses are slowly being offered on the internet. Even if your family is not strongly interested in science, a good basis in the subject is now more important than ever, for all intelligent families, and in my attitude, all families are intelligent.

In this book, the following chronology which varies with geographical regions, hence dates are to be considered very approximate. Stone Age Pre-4000 BC; Copper Age 4001-3600 BC; Bronze Age 3601-1200 BC; Iron Age 1201 BC to 400 AD. These dates vary from regions and lands around the world, and thus are highly variable.

The history of science from the early Asiatic, Arab, and Greece to modern times is not only an extremely fascinating subject, but can be used as an effective teaching tool for young and old students. If the old study science, they are able to teach and excite the young. The United States of America should remain the world leader in science. To maintain the status as a world leader, the young must be trained in science from early age. Such essential training of the youth can be accomplished by early reading facilitated by the gift of the internet, and public radio and television. The process of developing an understanding of science by the world's youth is vitally important, and should lead to solutions of current problems such as the development of new energy sources, economic studies, medical science discoveries, and what to do about the possible financial collapse of Greece and other countries.

I dream that early teaching of science for young children may lead to a new future for our world. We exist so that we may think, dream, invent, love and prosper. See how this important idea differs from Descartes' idea that "I exist because I think" (Cogito, ergo sum). Descartes who was born in 1650, was educated by his brilliant mother after the death of his father, before Descartes was seven. This is a great example of an intelligent woman who became the teacher of a 7-year-old boy, to be trained in science and philosophy. Never underestimate the value of young women as teachers for their children.

The Athenian Greeks developed a strong navy and a commercial shipping industry, from early times. The present capacity for ship building and merchant marine organizations based in southern Greece is located in an area called the Mediterranean Climate Zone, with warm winters, hot and dry summers, and cool nights—good for ship building, as compared with ship building areas located far to the north. The Southern California climate is also one of the five Mediterranean Climate Zones, along with Western-Central S. America, Western South Africa, and Western South Australia. Two Greek shipping companies developed the modern super-tanker ships. Cold, snowy, and rainy weather is uncommon in Greece, and that means ship construction is somewhat easier than in northern areas. As of 2012, Greek ship construction of large vessels is nearly at a standstill, with high unemployment of ship builders. With the infusion of European capital, the Greek economy might recover, when we do not know.

An age-old problem in Greece and many other countries was the lack of science training of young girls. With many girls spending their lives taking care of the home, they become teachers of their children, and therefore need math and science training to help the children to develop into future scientists, engineers, nurses, medical doctors, and better science training of future politicians. How can a world leader understand the economy and future engineering program development without a good background in science? How can political leaders make good economic decisions about new science developmental efforts, without understanding the importance of developmental science programs?

The age-old problem is that women are too often ignored in science training, but women are the best science teachers of children. Women rule the home, while the men are away at war or running businesses. Women become the teachers of children for the family. To do the best teaching, science training of girls is a vital effort. We exist that we may think, dream, invent, love and prosper. How could business prosper without 4000 years of scientific discovery, and girls as teachers, while their men were off to war? Cultures that suppress teaching of young girls in science will become much less prosperous.

Benjamin Franklin and Thomas Jefferson related good science to good government. Good science is important for good business and full employment. But how are your political leaders to know what is good science when many political leaders seem to misunderstand science? I think that our elected officials ought to be science trained along with parents and their young children. Our well-trained elected officials should never label attempts to develop good science as a myth. The ideas of Louis Agassiz's glaciations discoveries, Darwin's evolution, plate tectonics, sea floor spreading, and global warming have all been called myths in the early times to the late 20-th century. The idea that the Earth is moving about the Sun, and that the Solar system is not the center of the universe was considered not only a myth, but a heresy for thousands of years from the ancient Romans to Copernicus, Galileo, and Newton.

As of January 2012, there are many developments that can hinder science development and teaching. A highly important recent development is the modern television trend to non-science, and violent programs. Another example of a developmental impediment in the modern world is that more than a half a dozen states of the United States of America are enacting laws directing that teachers must instruct students in middle schools that students are to consider evolution subjects, and climate change ideas as false (not unlike the attack on Darwinism and evolution in schools in the last century). The net result of such laws may be that teachers will avoid trouble by not teaching current science when the subject might differ from ultra-conservative political or religious ideas. Even in this modern world, a minority of Arab groups are opposing the education of women. Their main interest is in forming a strong military. Such groups will bring on their own peril.

A further development is the attack on the internet in presenting current scientific ideas, interrupted by commercials. The result may be that the internet may become little more than a tool of the advertising industry. Wikipedia was shut down for a time in protest against pending legislation protecting the publishing industry. This author feels that the free internet should remain free, as a form of free speech. Many educators now are instructing students against using Wikipedia articles in school assays. The reason for this is to teach students how to think independently and to avoid plagiarism in school writing programs. But as Wikipedia evolves into an encyclopedia about every conceivable subject, is it wise for schools to prohibit use of references to Wikipedia and other such on-line articles, which should always include a reference list of available library books?

Another problem is the lack of good science programs on television, and the emphasis on non-science programs, even on public television. Good science programs on public television seem to be in decline. Parent-child reading sessions about current science developments have become more and more important. In November, 2012, KCET presented a science program, entitled 'History of Science'. Such a program was a good start. But it continued only for a few more hours, when the announcer stated that it might continue for more segments, some day in the future; we are still waiting. It represents a step towards a goal, but many such programs are needed. Encourage some of our best scientists to provide presentations on public television, and to develop science-teaching DVD's.

A major development at this time is the high cost of education. The very wealthy might be able to send their children to the best schools, but there are too few families that are super-rich and interested in science. The United States of America needs many more good students than can be supplied by wealthy families.

Many large universities have excellent lectures by faculty and visiting scholars. Such lectures should be recorded and put on public television stations, public broadcast stations, and on the internet.

A new and revolutionary approach to science as implemented by NASA, and Wayne Rosing, formerly from Google, is the evolution of internet-controlled, robotic astronomical telescopes for public use, using mass-produced 2-meter and smaller telescope arrays connected through the internet. These efforts and many other changes will revolutionize astronomy, as well as other sciences. Astronomers from many universities can now gain access to more and better telescopes, without having to fly around the world. At the same time high school students can get on the internet to carry out school-based astronomy projects without leaving the class-room. It is now possible to do day-time astronomy using internet connected robotic telescopes located where the weather is clear and the skies are dark.

Edwin Hubble and George Ellery Hale stated that light-gathering power was guiding the construction of the Mt. Wilson and Palomar telescopes. The light gathering power increases with the square of the objective, but the resolving power is limited by diffraction which increases at a smaller rate (the first power of the objective). In addition, the cost of a dome for a large telescope may increase faster than the cube of the objective diameter. Hence there is a limit to building larger aperture, ground-based telescopes. Smaller telescopes may actually produce sharp images, accessible to a larger group of scientists. Smaller telescopes are now being built, and are automated, and connected to the internet.

Reference www.lcogt.com

Future improvements utilize space-based observatories such as the Hubble in low Earth orbit, and deep space based telescopes located at stable Lagrange points distant from the Earth. Hubble required astronaut maintenance; reliability is clearly important for telescopes at the distant Lagrange points, expensive or nearly impossible to be serviced by astronauts. Life-time of full operations of a distant space telescope is limited to a few years, due to loss of cooling for infrared sensors, gyro systems repair, and radiation damage to the sensors and computer control systems. Cost remains as a serious limitation for space-based telescopes.

References:

Las Cumbres Observatory Global Telescope Network, LCOGT, started by Wayne Rosing.

The 2-meter world-wide telescope networks include the Faulkes, MacDonald, Bradford and other internet-connected and robotic astronomical observatories.

Sky and Telescope magazine, November 2012

Why is astronomy important? Astronomy is the oldest science, and highly important, historically. And most importantly for the survival of the human race, is the search for large objects that could destroy the Earth.

Simple star gazing by professionals as well as amateur astronomers could be the key to our survival by discovering new knowledge and possible detection of asteroids and comets about to strike the Earth. Suppose that a collision between a very large asteroid and a very large comet between Jupiter and Mars were to change the orbit of the collision products such that Earth is the target, how would we know and what to do about it. Perhaps the genius workers of JPL could be directed to intercept the debris of the cosmic collision with a high-speed, unmanned rocket to intercept the dangerous impact material, using the best ion engine technology, and with good trajectory computations, so as to deflect that monster projectile that might eliminate all humankind.

This book will use the internet in addition to the usual form of a reference section at the end of the book. The references for this book should also be available on a searchable CD. The references herein include references to articles on the internet. There are various ways to access the references to articles on the internet. One is to type in the full title preceded by the www, and so on. An easier way is to type in the item subject, letting Google or other search engines list the available articles on that subject. An experimental new idea is to make a typed copy of limited paragraphs from this book to a computer file, to allow rapid access to internet references directly on a page by page, and a subject by subject basis. Students should use the references given at the end of Wikipedia articles, rather than the article itself.

In order to simplify reference searches, the lists should begin with the primary author's name, with a cross-reference list by subject. Book indexes usually include author's names as well as subject names.

A word of caution about referencing any internet-related publication such as Wikipedia is that it may be difficult to verify the accuracy and completeness of such material, and it might be difficult to eliminate all advertisement and political inferences. Some educators do not allow use of a Wikipedia article in student papers. Wikipedia does, however, provide good references at the end of each article. It is best to reference one or more of those items rather than the Wikipedia article directly. Your personal library should incorporate listing of science references by subject, with help from Google.

A computerized list of references is included in the Appendix of this book. Occasionally, certain references may be out of print, or no longer available (even from Amazon.com or from Half.com). References on the internet may have been modified or removed. One should use Google or other search systems; enter the subject and review the list of available material (note that direct copying of the listings of articles by Google is to be avoided because such material is changed frequently and improved, and is to be used with caution.

In addition, enter the subject into Amazon.com to find the best available books on the subject. Go to a local public library for a copy of the books to be considered for reading or purchase. Visit a local university or college library. Get help from the library reference personnel. Print out the list of references found to be of interest. Study the material found for parent teaching of children, starting at an early age, during elementary and secondary school times to prepare for the future college time. Build a home reference library. A problem has been noticed, that many science books and magazine articles for the public leave out even simple math. This can cause trouble for both young and older readers. This problem could be reduced by adding math in a special appendix at the end of the book or magazine article in addition to the normal reference listing. Make it a routine to check with your library for important used books being discarded, or offered for sale at bargain prices.

How early can science and mathematics be introduced to young children? Elementary algebra can be introduced to bright students between the first and second grade, and should not be deferred to high school. The idea that science and mathematics teaching should be delayed until high school is strongly questioned. The best students in my grade school experience learned algebra by the second grade. A few students began to want to learn science by the fifth grade. A recent Nobel Prize recipient stated that he mastered the subject of calculus by the age of seven. I did it by the age of 18, in three weeks. The concepts of calculus are not so difficult that I could master the basic ideas in a few weeks, and a full calculus in 1.5 months, starting in the summer.

The first attempts to interest bright, young students should begin by visiting museums, national parks, botanical gardens and arboreta. For example, when my family took a trip to Los Angeles when I was 9 years old, before WW II, I found the Griffith Park Observatory science museum, and the Mt. Wilson Observatory the most exciting places to visit. Many large cities have such museums and botanical gardens, and should be a great experience for starting the education of our youth. Get help from the Museum personnel and library reference department on how best to find out what fields of science the student would find most interesting. And in the process of finding what the children find most interesting, find a way to encourage science study. Science teaching should be introduced early, and advertised as exciting, but should not be described as too boring or too terribly difficult. Avoid telling children that science subjects are too difficult for young children.

Recently, a national clothing store was selling a T-shirt for girls with a message saying, "I am too pretty to learn algebra". That overlooks that the girls may soon become home teachers of young children. For the United States of America to remain competitive, it is important, perhaps vital, that girls as well boys need to understand science. Our survival in the world may depend on the girls acting as teachers in the home. I believe that girls can make the best teachers as well as great nurses and doctors. It should become 'the American Way'.

Girls need to be introduced to science at an early age, so they can help the boys. Young girls frequently mature months (sometimes years) before boys; they are not drafted into the military, and they often live a few years longer than men. One reason they do not get drafted is they are needed to take care of their children, and grand children. Later as their children grow up and go to school, girls spend more time at home than men; the girls usually have more time to help children in math, science and history. If the United States of America is to compete with the huge population in Asia, then it is essential that women be trained to teach science and mathematics to young children. Early home education is an important way to minimize the cost of university study, which is rising faster than the cost of living, and the rate of increase in wages. Certain fields of study are much more costly than others. For example, a medical degree may require more than 12 years plus a year of internship. A good scholarship would be helpful. A good way to get a scholarship is to prepare years ahead of time by early-age home study, and in-home, study groups.

The young are fast learners. The young love new ideas. Why is it that the young often understand computers and how to access the internet better than their parents? Computer training could be on TV.

I envision a dream: summer, night-time and weekend home discussion groups led by groups of parents and young children, with a bit of guidance by a few retired teachers and scientists, plus good science programs on public FM and television stations. Groups can meet in homes, churches and library meeting rooms. First take the young students to local museums. Discover what interests each student the most. Then schedule a continuing program of discussion with the young students with guidance from current and retired professionals. The idea is to develop the young for study and reading with help from library experts. Start the home study program well before grade school, and continue through high school and into college. Excite our youth by introducing mathematics, science and history in a family-based learning program. Home study should become an essential augmentation to early standard school curricula.

Such group sessions including parents, friends, and young children ages from 5 to 18, or earlier, might make a great program on local, independent TV stations. Demonstrate how local groups can utilize such parent-student-child discussion-learning sessions to help develop students in the great world of science. Invite present and retired teachers and scientists to discuss specific scientific subjects. Contact colleges and scientific organizations for help. It is very important that our best scientists give help in developing better understanding of science, and related fields such as medicine, history, physics, astronomy, biology, geology and economics. Such a program is to augment regular classes, private schools, and tutoring agencies by using family-oriented reading and study efforts during evenings, weekends and summer vacations.

Bright children come from all races and cultures. It appears that learning foreign languages at early ages leads to development of brain power in children 3 to 6 years of age. That may be one reason that Chinese and Asiatic children coming to the United States of America are among the brightest students. Students of Jewish and European backgrounds are also included in that group of students with multiple language skills which may be why family-oriented language use may produce outstanding abilities in the fields of science. If multiple language use in the family is of importance, why not include math and science discussions in early-age, family discussion sessions.

Not all children are equally gifted. Some are born gifted, but the vast majority of children simply develop brain power if exposed to subjects that are interesting to them. With careful guidance from parents, schools, libraries and museums, the average student may become gifted through good family-based teaching and encouragement. Learning science at an early age leads to good brain development.

If it is true, that learning a second language before the age of four is easy, and helps to develop brain power, then we should offer second-language courses beginning in kindergarten or earlier. My son had no difficulty in learning English and German before the age of four. He never mixed up his sentence structure, or had trouble using English. He is now learning Spanish.

A short list of science-related subjects that could be taught by parental help at the fifth-grade school year for helping in the development of bright students include mathematics, physics, science history, biology and medicine, plant studies, geography, geology and Earth science, economic history, space science and astronomy, agriculture, animal studies, computer science, and human evolution.

Many magazines leave out even simple math formulations when discussing science, even Scientific American seems to avoid math. It is important that at least the basic math be included. The basic equations not included in the text, should be given in an appendix to each science article to help educate the public, our children, our political leaders, our teachers, and parents who may not have been educated in the subject of the article.

I went to an all-male, high-cost, prep school, and a top-level university. Some were all-male institutions. They reasoned that girls were destined just to populate the Earth with children, and high-cost colleges need not waste their resources on girls. I had several brilliant girl friends who could never get advanced science education. Most of such all-male schools have changed, and now admit girls. I was born too soon. In my day, the most beautiful and intelligent girls dated and married local sports heroes—why were they not dating and helping to develop future scientists and medical doctors? Times are changing.

Girls may find that they can obtain early and advanced science training in many cultures, and can relate science education to future family prosperity.

Enough, said, let us return to the struggle of early scientists, and learn how the ancient Asiatic, Arab and Greeks attempted successfully to carry out such programs of study. Use the ideas of the ancients to teach our children by parent-oriented study sessions. Mathematics and science in ancient Greece was taught by visiting lecturers given by the Sophists. We now call such instructors, tutors, which in our time period should include parents in parent-child discussion programs.

This history of science begins with the early Greek as well as early Asian and Arabian scientists who lead the way for measurements needed to describe the Earth and surroundings. Science is based on measurements, and those measurements required accurate definitions of units, and procedures to be followed. The early Greeks lead the way. World wars and diseases caused much suffering and delay. In ancient Greece, many young scientists began to develop their ideas after their service in the military. It is not the purpose of this book to dwell on the brilliant military leaders of Greece and Rome and their attempts to control and tax the world populations, but to illuminate the step-by-step lengthy process in development of scientific thought and discovery.

The fight that it frequently took to convince doubters of scientific ideas when those ideas differed from common thinking and religious dogma will be covered in at least some detail in this book. The fear of falling off the edge of a flat Earth was happily eliminated, but the idea that the Earth was not the center of the universe was attacked for thousands of years up to the time of Copernicus and Newton, and resulted in preventing future discoveries by Galileo by his house arrest and prohibition of his scientific publications, just because the observations through his telescope showed that the Sun and Moon were spotted and not a perfect product of God and his perfect works.

Such attacks against scientists have continued to the modern era, whenever ideas conflicted with common thought. It is as if freedom of speech did not apply to scientists.

Major conflicts such as world wars I and II, clearly had an effect on scientific development. In the present dangerous world, where isolated groups of individuals can blow up and kill random groups of people, and, further, groups of wealthy individuals can spend large sums of money to impose their ideas, using large groups of ultra-conservative people to get them to attack scientists who differ with the ideological positions of extreme conservative thought related to current scientific thinking. I should not bring up the subject of present-day conflicts in the middle-east, but such troubles can have long-term impact on the advancement of science in such countries. Such subjects are beyond the scope of this book.

Lists of the world's scientists and discoveries given to us may be presented in an appendix and on a computer file that can be easily searched and quickly accessed by readers. Lists in the Appendix will also provide time lines for major wars and diseases. If this material were limited to a pile of printed pages, access would be somewhat restrictive. The idea is that if this book or at least portions were included on an electronic medium, then such time lines, listings of scientists, and discoveries can be accessed by the availability of fast search engines. Those search engines should be extended beyond the material in the appendix to include references on the internet. The details of the search algorithms and methods of implementation are beyond the scope of this book. One may, however, consider that capability is readily available. Get help from library and school guidance counselors.

The costs of education for our bright, young students are increasing much faster than the cost of living. University tuition and student housing cost increases could make education past high school only for the super-rich or a few super-bright kids. One way to avoid the high costs of advanced education would be to start the scientific training of our youths by early home study sessions with help from the internet. If important science and mathematics were introduced as early as the fifth grade or even earlier, then more of our bright, young students could qualify for scholarships to college. This is the American way.

The introduction of science should become an important task for summer evenings. Groups of children and parents could be gathered together for home-community groups using computers and the internet with guidance from our best teachers, retired scientists and with help from libraries. This book is structured toward such a program. There is not a more important educational program for our young than early scientific training. This history of science should also be an exciting and important effort for parents reading to children. Our future world needs better scientific understanding. I hope that home-based, family-oriented reading and discussions might lead to better understanding of basic science in the United States of America, and in other countries as well. Perhaps, home reading and family-based discussion sessions might make the US a far stronger nation.

A troublesome future problem exists: what to do when the world runs out of low-cost oil and gas this century. Do we turn to coal with its dangerous contaminants? Even coal will eventually run out or become very costly. We need a new set of energy sources. Future research may provide the answers for human survival. Early education of our children via increased family reading is an essential goal, instead of so many unimportant and/or intellectually boring TV programs. What should we to do if a large asteroid or comet is found heading to Earth? Ask your local public television stations to provide more NOVA science programs.

I would also suggest that children's book authors consider writing science books for young children.

In the US, the standard set of units is called the pound-foot-horsepower-second system. Beginning science students are introduced in the pound-foot-watt-second system. Only the United States of America, Burma (or Myanmar), and Liberia use this system. Early science students are taught using this system. Later these US students are taught in what is called the US or Engineering system of units. Then later students are taught the metric system. As a result, some confusion exists as a result of trying to teach two or three different systems. Engineering students are taught in a hybrid of the metric and engineering units, with some confusion as a result of teaching in these systems. Many engineering students get a mixed idea of fundamental units. The following is an attempt to reduce the confusion.

In the US, units include the pound, foot, inches, miles, with power in horsepower, and the second. And the relation between pounds and the kilogram, is approximately 2.055 pounds per kilogram at sea level (a mixed relation between the US unit system, and the metric system). The pound is a unit of force or weight, at sea level. The kilogram or gram is a unit of mass, or quantity. In the US, the slug is the mass unit.

HISTORY OF SCIENCE

Science and invention began with the early Asiatic, Chinese and Arab cultures. The Chinese invented silk, porcelain utensils, the magnetic compass, an early seismograph, sailing boats, rice planting, writing paper, and a form of printing using Chinese symbols carved on blocks, gun powder, the cross bow, domestication of farm animals, and astronomy with star catalogs by 400 BCE. The Chinese introduced the decimal system with powers of ten, and paper money. The Chinese imported bronze, iron, steel, and chariots from the Arabian cultures in the Mediterranean regions. Rice was developed in the Indus Valley located to the northwest of India. Early science and agriculture started by Arabs in Mesopotamia.

References:

The Genius of China, 3000 years of science, discovery and invention, by Robert Temple, 1986

Ancient Chinese Science, by Robert Silverberg (a book for the young)

History of Astronomy from Ancient times to the present, by H. Couper and N. Henbest

INTRODUCTION TO SCIENCE FOR TEACHING YOUNG CHILDREN

An introduction to the science of the Earth as a system for teaching both young and mature adults may begin with tables of fundamental units and discussions. In ancient times, measurement systems varied with location and time period. Each country had its own set of units. For example, Eratosthenes, in 200 **BCE,** measured the polar diameter of Earth accurately from distances and sun angles along a path adjacent to the Nile from tropical Africa to Alexandria in Egypt. He measured distances in stadia. But we do not know what value of stadia he used. If he used the Egyptian stadia (157.5 meter per stadia), his result was within 2 percent compared to present measurements for the polar circumference of the Earth. A later scientist in Egypt measured the equatorial circumference of the Earth, but in error by more than 3000 miles. That error led to Columbus ending his sail to the Americas thinking he had reached India.

Units and measurement systems must be stated in order to find agreement between ancient and modern scientists, and in communication with others in foreign lands. It should be remembered that pounds and ounces are used primarily in the United States of America, Myanmar (Burma), and Liberia.

The pound as used in the United States of America is a unit of force (the weight due to gravity on an object at an equatorial latitude). In the United

States of America, the unit of mass is the slug, which is not in common use, even in schools.

Mass Versus Weight

The difference between mass and weight is very important. Mass is the quantity of a material and can be measured in kilograms. Weight is the local force of gravity on an amount of material measured in slugs in the United States of America. The pound is a unit of force. Mass in slugs is a unit of quantity.

A small quantity of gold may be stated in terms of the Troy Ounce, but that measure is not actually a mass unit. Avoid using mass and weight interchangeably, except at the grocery store and your weight measured in a doctor's office in the US.

The pound at this time in the US is used in the kitchen, the grocery store, and on bathroom scales.

Until reliable springs and hardened coiled steel springs were produced all quantity determinations were made using calibrated mass standards and precision measuring cups. The pound measurement required precision spring scales, not available until the middle of the 18th century, at the time of Hooke who worked on the physics of coiled springs.

A minor problem with using the pound as a mass unit, is the dependence on altitude. Is your bathroom weight measurement valid for your altitude? Is your doctor's weight scale properly calibrated for altitude? Most bathroom scales have an adjustment dial. That dial can be mistakenly set incorrectly. Does it matter that much? An accidental readjustment can mimic a weight gain or loss. To avoid an error, provide a standard kilogram weight to check on the calibration of the scale. The pound scale in the grocery store can be inaccurate, but good enough for even an expensive steak.

The spice trade required an agreement on quantity, mass and weight units. Precious metals and gems also required agreement on quantity and weight units. The equal pan balance was developed in early times in the Indus Valley, Mesopotamia, and Egypt. The equal pan balance required agreement on a mass standard. An early mass standard was a set of bronze mass units. Another standard was a measured quantity of pure water. The addition of a calibrated slider to an equal pan balance was developed. For precious metal or gem stones, mass measurements with equal pan balances were developed using a set of standard mass units.

Mass and weight are not interchangeable. The relation of 2.205 pounds per kilogram applies to sea level measurements. Those troublesome pounds relations are inexact for measurements at high altitude. Children should be introduced to the kilogram and other metric units as early as preschool time. The US pound is a mixed up unit. It is used incorrectly as both a force and a mass unit in the US. Use of the pound as a mass unit leads to confusion, and lots of errors.

The early development of the Scientific Method

Socrates introduced a method of teaching called elenchus, which is a procedure for a class or a group by posing a question, followed by group discussions, leading to an hypothesis. The hypothesis then is subjected to positive and negative consideration by the class, and is to lead to a refinement. The next step is experimentation, or an analysis of and-if ideas of a thought experiment (the German Gedanken Experiment). The next step is the formulation of a theory to be tested by experimental measurements, and a process of peer review. Socrates' idea is often used as a teaching method in law schools. Much of what we know of Socrates comes from one of his great students, Plato. Later statement of a scientific method was carried on by Galileo. Every great scientist in antiquity tended to develop a guide to his central idea development. These guides became their personal scientific method.

Examples of difficult problems in early science idea development include: is the Earth a fixed center of the universe? Do the Earth and Sun move, and by how much, and in what orbit? Was there an ice age, with thick glaciers covering the northern lands of Europe, Asia, Canada and North America? What causes the climate to vary with distance from the equator, and with seasons, and how it varied from present to ancient times? What causes the tides? Can industrialization lead to climate change? How can volcanic eruptions change the climate? Can extinct forms of life found in rocks be used to date the age of the rock strata? Considering rock classifications of sedimentary, igneous and metamorphic, what geological process led to metamorphic rocks? What herbs are needed for good nutrition? What is the evolution of plants and animals? These are just a few examples of human thought and heated arguments lasting hundreds to thousands of years. Possible future problems may include how to keep your house warm or where will electricity come from, a couple hundred years from now when the Earth could have run out of coal, oil and gas? Do not ask the Tea Party Republicans, as they have been taught that the Earth and the United States of America have unlimited resources, and making great changes in our fuel use would be much too costly, and lead to loss of jobs, and reduced profits of certain corporations that provide financial basis for the Republican Party.

The Greek scientist named Democratus, in 400 BCE, reasoned the all matter was composed of tiny atoms. What were these atoms was not known for at least another 1500 years. Early thinkers knew that lead could not be changed into gold or silver. Alchemists would try to do that for many centuries. Even Isaac Newton in the 1600's would try to do it. Such attempts could lead eventually to the idea that atoms that make up gold or silver are elements that could not be constructed from base materials, such as lead or iron, or changed in a process called transmutation.

A radical change in thinking was introduced by Irene Curie, by the discovery that certain small changes can be made by irradiating an element with alpha particles and neutrons. It might be possible to change gold to a different elemental form with a slightly different atomic mass number.

Further advancement in scientific methods was introduced by Galileo Galilei. Reference the many books on Galileo. Galileo's ideas conflicted with the Catholic Church dogma, which was largely based on the ideas of the ancient Greek scientist, Aristotle. This is a prime example of serious conflicts faced by scientists who presented discoveries that differed from current thinking and ideas held by various religious leaders.

References:

Galileo, the Man, his Works, his Misfortunes, by James Brodrick, S.J., Harper and Row, 1964

S.J. Refers to the Society of Jesus. James Brodrick was a Jesuit priest.

Galileo, by Colin A. Ronan, G.P. Putnam's Sons, 1974

Reference various encyclopedia volumes. Compare what books tell you with Wikipedia articles.

Galileo initiated an effort to standardize the fundamental units, an essential step in allowing comparisons between different authors in different parts of the world. Rather than giving his lists which could lead to confusion, current definitions of selected units and measurements are given below; further details are given in Appendix U.

MKS System and Mass Versus Pound Units

The mass standard is the kilogram by international agreement, and is based on the standard kilogram in a Paris science depository. The pound is a unit of weight or force, and is official only in about three countries (USA, Myanmar and Liberia). The pound is a measure of the weight of an object on the surface of the Earth. The pound is not a mass unit. The pound is related to the kilogram by the approximate ratio of 2.205 pounds per kilogram on the surface of the Earth, only near sea level. The sea level and terrestrial altitudes vary over the Earth, hence the pounds per kilogram ratio can vary. If one weighs an object on a equal pan balance, using a standard pound set of weight units, few problems are encountered. But if the weight of an object is measured on a spring scale, problems can occur. The weight of an object in pounds varies with altitude, and from place to place over the surface of the Earth. If one orders a pound of ground beef in Washington, DC there are no difficulties. But one may get more or less beef per pound if purchased in Denver, Colorado using a spring scale, unless the scales are readjusted properly for the mile height of Denver.

Stability of the Platinum-Iridium kilogram mass standard

All Pt/Ir mass standards gain mass slightly, but measurably over time. A carbon-12 and a silicon-28 mass standard have been suggested, for consideration in the year 2017. The silicon crystal-growing technology is highly advanced, and a spherical kilogram mass standard of ultra pure Silicon-28 has been demonstrated. A mass of 27.9769265325 of Silicon times 6.02214179 x 10 raised to the 23 power, giving 35.74374043 moles of Silcon-28 has been produced and is proposed for a new standard kilogram.

When using Newton's equations with pounds, the acceleration of gravity must be put into the equations. In weightless conditions in Earth orbit, the weight in pounds is zero. But the mass in kilograms stays constant around the Earth as well as in space. Work in the American system is in terms of foot-pounds, or in the scientific MKS system, newton-meters.

The Newton's equation relating mass, m, and force to acceleration given in the metric system is

$F = m\, a$

where the mass is given in kilograms and acceleration in meters/sec/sec, and the force is in units of newtons. The force is usually used for measurements in the gravitational field of the Earth and in the Solar-interplanetary system. For operations of a person pushing a wagon, the force involves both gravitational and frictional as well as other non-specified force fields. One may measure the mass of a loaded wagon and the acceleration, but the force fields normally may be a mixture of friction and other force fields. It might be difficult to separate the push as well as the impulse. Most measurements need to consider the motion as a result of impulse and momentum. For example, the behavior of a golf ball depends on both gravitation and air frictional forces, and friction between the turf and the ball.

For the American system of units, the Newton's equation equivalent is: the force F, in pounds, is given by (valid at sea level):

F (pounds) = Weight (pounds) times the acceleration (in feet/sec) divided by the gravitational constant, g. In proportions, W is to a, as F is to g. The value of g, in these relations is 32.174.

Where g = 9.80663 meters/sec/sec/2.54 centimeters/inch/12 inches/foot = 32.174 feet/sec/sec

A few relations are as follows. In the US, the following formulas are to be used:

Mass for USA = Weight (in pounds)/g (the acceleration of gravity 32.174 feet/sec-squared)

F(force in pounds) = Weight(pounds)/(a times g acceleration of gravity in feet/sec^2) or F=aW/g

Kinetic Energy (foot-pounds/sec^2) = weight(pounds)/(g acceleration of gravity) x velocity^2)/2

There is a series of standard units used in the US. This system of units is referred as the pound, foot, second system. The comparable metric system is referred to as the kilogram, meter, second system, the MKS system, for short. When writing a science article, intended for readers outside North America, use the MKS system of units, to avoid problems.

If a rocket has a measured thrust in pounds, it must be converted to metric units, for NASA or international work. The approximate conversion for pounds thrust into metric system units is 9.8/2.205 = 4.448 newtons per pound, at sea level. Note that the unit 'newtons' is not to be capitalized.

A few standard MKS units are listed and described in Table 1.1.

Table 1.1 Standard Units and Conversions (see the Appendix for a full listing of important units)

The MKS System of Units

Kilogram -- a mass unit is based on a standard in an environmentally controlled vault in the Pavillon de Breteuil in Sevres near Paris, France. Current standardization of units started prior to the French revolution by Lavoisier, considered the father of modern chemistry. The values of units were adjusted according to new information and discoveries, and documented every few years by international agreement of scientists. In 2016, the definition of the kilogram in terms of the Planck constant instead of the standard kept in the vault in France might be considered.

The current MKS (meter, kilogram, second) called the rationalized MKS system, replaced the older CGS (centimeter, gram, second) system in an earlier-year time period. The rationalization involved improvements in the electromagnetic constants and the CGS system constants. Future agreements might consider defining the kilogram in terms of the Planck constant instead of the reference standard of mass in the vault in Paris.

A listing of important fundamental constants include the speed of light, Newton's gravitational constant, the standard acceleration of gravity which varies over the surface of the Earth, electromagnetic constants, the Planck constant, stellar and planetary constants, and the geological time scale tables which are given in the Appendix for units.

After reading, the following may be skipped over, but not forgotten. Some vital unit conversions should be memorized. To get a feel of the metric system, a 32-ounce bottle of Gatorade refilled to the top with tap water has a mass of about 1.05 kilograms, including the plastic bottle. The label indicates that the contents of the Gatorade 1 quart bottle has a mass of 0.946 kilograms (32 fluid ounces, 1 quart, 946 ml). A ream of 8.5 x 11 inch, 20 pound bond paper weighs 5.05 pounds, approximately.

This is a trivial matter. Let us have a little fun discussing singular versus the plural form of units. The professionals who constructed the MKS unit systems suggested using the singular form in referring to standard units. Data are plural; datum is singular; most measurements involve many datum points. Meters is plural. Feet are plural. I suggest that the use of the plural form of units sounds better than the singular. Just by use of sentence construction one can avoid problems of plurality. For example, we can say that the equatorial circumference of the Earth is 40075.04 kilometers, or 24,901.45 miles, or 131,480 feet (not 131,480 foot). Thus one should say that the Egyptian stadia was about 157.5 meters in length. Use the plural form of the unit name, for less confusion, when constructing a sentence for a book to be read by Americans. For European publishing house editors, follow the usage dictated by the General Conference on Units Weights and Measurements in London, England or Paris, France scheduled for possible changes in 2016, when writing for technical journals.

The conversions from feet to meters and to kilometers, based on 2.54 centimeters per inch, are:

Multiply measurements of length in inches by exactly 2.54 to give centimeters

12 inch/foot x 2.54 cm/inch = 30.48 centimeters per foot

100 centimeters/2.54 centimeters/inch = 39.37007874 inches per meter

100/30.48 = 3.280839895 feet per meter

100,000/30.48 = 3280.839895 feet per kilometer

5280 feet/mile/(100,000/30.48) = 1.609344 kilometers per mile; 60 miles per hour is 88 feet/second. (60 times 5280 feet/3600 seconds) Are these relations correct?

A selected list of units, formulas and conversions are given below in Table 1.2

Table 1.2 Selected Fundamental Units for a more complete listing see Appendix U

Speed of light in a vacuum; symbol c = 299792458 meters/second or 2.99792458×10^8 m/s

Constant of gravitation, symbol G = 667300000000 or 6.673×10^{11} meters/(kilogram x second^2)

Formula for classical (Newton's) gravitational force is $F = G(M_1 M_2)/R^2$

Equation of Earth's elliptical orbit is $(x/a)^2 + (y/b)^2 = 1$ and the eccentricity, ecc = $(1-(b/a)^2)^{1/2}$

Consult various reference books and the internet for the values of constants in the above equations.

Earth perihelion distance = 1.471×10^8 ; aphelion = 1.521×10^8 ; The Earth orbital eccentricity = 0.017.

Earth mass (kilogram) = 5.97×10^{24} Mass of the Sun (kilogram) = 1.98892×10^{30}

Average distance of Earth from the Sun = 1.496×10^8 kilometer = $0.5 \times (1.471 + 1.521) \times 10^8$.

Relations between distance traveled, velocity, acceleration and time:

Velocity, v. acceleration, a and time, t v = at

Distance, s s = 0.5 a t-squared

Ancient Units in Greek and Roman Time period:

The Greek Foot was 0.3082 meters (compare with the present value of the foot = 2.54 x 12 = 30.48 centimeters per foot, or 0.3048 meters per foot).

Cubit 0.4623 meters (the average length from an adult human elbow to finger tips).

Roman Mile 1379 meters

Talent 37.8 kilograms

For example (Noah's Ark, Genesis 6-9, referencing early times perhaps as early as 7000 years BCE):

Length 300 cubits; 139 meters or 455 feet

Width 50 cubits; 23 meters or 75.8 feet

Height 30 cubits; 14 meters or 45.5 feet (It was a ship of large size for that biblical time)

The Bible, Genesis, chapters 9-21

Such a ship must have taken longer than a few weeks to build. At the very least, it could have taken an entire life-time of a generation of a ship building family. It was more than a barge. It had three decks. A boat 300 cubits in length would have been a major project. That project appears to have survived today, buried under many layers of sand and dirt. The site of a buried ancient ship has been found, and has been studied by scientists who feel that it must be the remains of Noah's Ark, as it has the right dimensions. And the wooden remains have turned to stone appropriate for the age.

Now, the Noah's flood was not the first time a major flood was described. The older legend of the Gilgamesh may have been the first. A huge flood was not the only major event that impacted early civilizations of the past. For example, there was a major super-volcanic eruption about 70,000 years ago. That eruption was located at the present day Lake Toba. Lake Toba is a large volcanic caldera near Sumatra, Indonesia. There is no written literature on the effects such a large volcanic eruption could have caused—such as a major flood. Think about it. A super-volcanic eruption near an ocean in southeast Asia would have resulted in a very large tsunami, with multiple wave inundations with wave heights exceeding 300 feet for certain coastal waters. Further thought on a biblical flood may have resulted from word-of-mouth stories about flooding which could have resulted from a sudden breakup of northern continental ice sheets. The last ice age breakup began about 12,000 years ago. No mention of the ice covering the northern regions is mentioned in the Bible. Could a sudden breakup of glacial ice have caused a biblical flood? Could major eruptions of the volcanic island called Thera, or Santorini south of Greece caused a massive flood of the entire seas and coasts of the middle east?

A more recent flooding in the Mediterranean sea about 4000 years ago (about 1600 BCE) wiped out most of the pre-Greek Minoan civilization on Crete as a result of the massive eruption and collapse of the volcano called Santorini or Thera, and the subsequent massive tsunami with possible wave heights exceeding 300 feet.

The Ark as a large ship would have to have been constructed near a waterway or ocean inlet. The time period of 70,000 years was much earlier than Noah's time, but nevertheless it could have become the source of legends of severe flooding. The explosion of Thera in the Mediterranean around 1600 BCE could also have been a more recent Noah's flood. This is just a matter of speculation, and is not readily verifiable material, as remnants of Noah's Ark which may have been found. The site is protected by the Turkish government. What local wood was the Ark made of (the Atlas cedar (Cedrus atlantica), or a Pine such as the Aleppo Pine (Pinus halepensis), or cypress (Cupressus sempervirens) growing near the Mediterranean Sea. The English bible states that Noah was to construct the Ark from gopher wood. What gopher wood was is not known. Would marine boring fishes and other inhabitants of the ocean have devoured the remains of the Ark? The wood of the Ark could have rotted, and turned to stone.

There is substantial evidence that the buried remains of a boat with the dimensions of the Ark have been found in 1985 in an area in Turkey near Mt. Ararat. The Turkish government has established a visitor's center at the site. The area has been surveyed with modern instruments, and the remains are clearly a ship matching the Biblical description of the Ark, buried on an area, some distance from the two volcanic mountains. The last eruption was about 1850. Wood samples have not been taken from the site, as far as I know. The site has been left, undisturbed.

References: a 1958 American television presentation.

The Encyclopedia of Geology

For articles on the Toba and Thera eruptions, refer to articles on the internet.

The Toba Catastrophe Theory on the internet (Wikipedia.com)

References:

Dave.williams@nasa.gov (NASA Goddard Space Flight Center lists of units)

The Chemical Rubber Handbook section on units and conversions

The US Nautical Almanac, published yearly, until put on the internet

Menzel, Donald H, and Pasachoff, Jay M., "Stars and Planets", a Peterson Field Guides Series, second edition, Houghton Mifflin, and later field guides than 1983

Geologists relate Earth strata by a set of time periods. The geological time periods were originated by early scientists in England. The originators were William 'Strata' Smith in 1790, James Hutton, and Charles Lyell. Lyell's book was used by Darwin during his voyage on the Beagle. A more modern but simplified version is given to geological students. A more complex table in full color is available on the internet, see stratigraphy.

Reference:

International Committee on Stratigraphy

Very early geologists thought that all rocks were sedimentary, and had been laid down under a warm ocean. More recently, the ocean floor was found to be almost entirely basaltic with a more recent age than many continental mountains such as those near the east part of the United States and the European western Alps. The Alpine strata had been folded and overturned, with many older sedimentary and metamorphic rocks on top. Alpine geology is complex, and had been studied extensively in the 1800's. The mountains along the western coast of North and South America are younger than 15 to 30 million years.

Mount Blanc between France and Italy is the highest Alpine mountain, and is composed of very old granitic and metamorphic rocks, lying on a sedimentary base. It has been described as a mountain without roots. Detailed geological rock descriptions for Mt. Blanc are complex and rather hard to find and understand. How could the Alps have been formed without considering plate tectonics?

Biblical scholars calculated the age of the Earth from ancient texts and the bible to be as late as 7000 BCE. Isaac Newton calculated the age as 16,000 years, based on time it took for deposition of rocks from a space filled with dust and other materials. Newton's calculation did not include any contribution from radioactive materials, unknown before 1900 AD, thus the time it required for cooling of surface deposits was greatly underestimated.

Reference:

International Commission on Stratigraphy

The Oldest Rocks on the Earth

Only within the past 50 years has there been any accurate and repeatable ways to determine the age of a rock. The oldest rocks are meteorites (some are between 3 and 5 billion years old). Prior to the Apollo Lunar Landing, astronauts were trained by studying old granitic rocks found in the mountains near the summit along Route 2 to Palmdale, California. These rocks are called anorthoclase, about 2 million years old. Similar rocks were found on the Moon. As a result of Apollo, Lunar rocks were returned to Earth and studied thoroughly. These rocks were about 4 billion years old, helping to prove that the Moon had been ripped out of the Earth by a massive meteor, during the earliest forming stages of the Earth.

Meteor and Meteorites

The meteor is applied to an object in space, prior to Earth impact. A meteorite is applied to the material after striking the Earth. The study of meteorology has nothing to do with meteors, only weather phenomena.

Meteorites are classified into several categories (iron-nickel meteorites, carbon-chondrites, and combined meteoritic materials). Most meteors striking the Earth fall into the sea. A few fall into snow and ice fields in the Arctic and Antarctic regions. Almost none are found along sea coasts. Many are found on inland areas.

References:

1. The Chemical Rubber Handbook lists of fundamental units
2. The US Naval Almanac, published yearly
3. Various recent and not older than 1970 geological text books
4. Sky and Telescope magazine, some issues, only
5. Menzel, Donald H. and Pasachoff, Jay M., "Stars and Planets, Peterson Field Guide Series", Houghton, Mifflin, and later publications older than my 1983 issue
6. International Commission on Stratigraphy
7. Scientific American

A few important relationships relating velocity, distance traveled, acceleration versus time are:

Momentum, p = mv

Page 11 Chapter 1 book page 27

Velocity, v = a t, for acceleration and time

Distance s = 0.5 a t^2

Work in the American system is measured in foot-pounds.

1 liter of pure water at +4 degrees C weighs 2.204684 US pounds

Power is measured in foot-pounds/second.

Horse-power is about 550 foot-pounds/second.

Power in the British system and in the electrical system is in watts

An example for a modern car:

How many seconds does it take to reach 60 miles per hour, at full and steady acceleration?

It should take about 6 to 10 seconds for a car. The average acceleration would be about 10 miles per second per second. Testing your car is a poorly-controlled experiment, and possibly somewhat dangerous. It is poorly-controlled due to the fact that the acceleration varies during gear shifting, during which time the acceleration drops momentarily to zero, even with an automatic transmission. An easier experiment is for a falling object. Read about Galileo's experiment for a lead ball and a block of wood dropped from the Leaning Tower of Pisa. He observed that both objects fall at the same rate. The rate of acceleration of a freely-falling object, neglecting air friction, is steady, and about 32.174 feet per second per second.

One could do such an experiment by filming the fall of a spinning flashlight dropped from a balloon on a windless night at a safe, desert location. (32.174 is about 9.8066 centimeters per second per second). A falling object has a more nearly steady rate of acceleration than an automobile. During daytime, a spinning mirror could be used instead of a flashlight.

For a falling body, the distance of fall is given by the formula of $s = g$ multiplied by time-squared divided by 2, where the acceleration of gravity is g (which decreases somewhat with altitude).

Take caution not to stand near the impact spot. Fasten a string to the balloon and another to the flashlight release device so that the flashlight and balloon would not go too far astray; and in the event of a failure of the release device, pull the string and all materials back down to Earth, safely. If this experiment is done from a manned, hot air balloon flight, there may be a safety problem. How should one keep people away from the drop site? All experiments involve safety concerns, which must be addressed. A safe and useful experiment needs some form of peer review before, during and after for data analysis and presentation operations.

Other Experiments for Youth Demonstration

References—use the internet for science experiments for children

Rate of motion of a smooth marble or a rolling AA battery, down an inclined plane.

Pendulum motions

The period of a pendulum, T, is equal to 2 times pi times the square root of the length of the support divided by the constant of gravity, g. The pendulum is a way to measure local gravity.

Stream flow rates

Search for magnetic rocks

Density and stiffness of engineering materials (steel, aluminum, wood)

The Chemical Elements in ancient time

The development of the concept of the elements started in ancient times. The early concept was that there were four elements (Earth, Air, Fire and Water). The idea of the elements is that all materials and substances are made up of fundamental and individual atoms. For example, gold was found in rocks. Gold can't be subdivided into smaller units—it is an element. Materials known from early times included wood, stone and rock, salt, water, gold, and Earth. Iron was known from certain meteorites. All materials were thought to be made up of smaller items called atoms. The idea of atoms was a concept that took thousands of years to develop.

Notes

page 13 Chapter 1 book page 29

With a distance measurement using a GPS reading on a Samsung cell phone, good to 0.1 arc seconds, about 10 feet.

0.1 arc sec times 24901 miles times 5280 feet/mile divided by 360 arc sec per degree, divided by 10 equals: 24901*5280 feet/360 sec per degree/3600 arc sec per degree-sec/10 = 10 feet.

To test this formula, find a street that is straight and heads to the east, then make a series of GPS measurements along a path measured with a 30 to 300 foot long tape. Take GPS readings at the beginning of the tape and again, farther to the west. Then take a second set of measurements farther from the starting point, at intervals of 1 foot up to 10 feet. Note how far it is before the tenth-second last number to the west or to the north, when the last digit increases by 0.1. Determine how far that takes for an increase. Repeat readings, east and west, to determine the accuracy. Note the variations in the repeated readings. It is expected to average 10 arc seconds or better for 10 feet per arc second using a simple GPS unit. The military version of GPS, or the high-precision and high-cost survey GPS unit is precise to about 0.1 inch.

ANCIENT SCIENCE AND MATHEMATICS

The Babylonians in early times developed mathematics and science, and shared their discoveries with the Greeks. Algebra was developed earlier than the classic Greek period. Early Greek thinkers learned astronomy from the Babylonians. Early scientists included Thales of Miletus (624-560 BCE), and his pupils, Anaximander (610-545 BCE), Anaximenes (570-500 BCE), and Diogenes of Apollonia (by 600 BCE). Pythagoras (570-495 BCE) studied in Babylon where he learned the triangle formula. Marco Polo (1254-1324 BCE) and Genghis Kahn (1162-1227 BCE) might have noted the early developments of math and science. What we know of early algebra is in a translation of Arabic material into Latin, Liber Algebrae et Almucabala, by an unknown western author. The ideas of beginning algebra can be found by reading Algebra for Dummies and other such beginner's books. Near the end of this chapter, a limited history beyond the time of 100 AD is included and such items may overlap with those in Chapter 3.

Reference

Early Greek Science

Thales of Miletus (624-546 BCE)

Thales developed some of the early ideas of plane geometry. He correctly predicted the solar eclipse in 585 BCE.

Anaximenes (535 BCE) and Anaxagoras (500-428 BCE)

Little is available in my local library on these men.

Pythagoras (570-495 BCE) p 30-32

Pythagoras may not have originated the law of right triangles, which he obtained from the travels into Babylon. Pythagoras, however, was famous for developing a proof of the triangle formula.

Pythagoras is known by his formulation of right triangles: the famous and very useful $C^2 = A^2 + B^2$ formula. The symbol C is the long side or hypotenuse, A is one side, and B is the other side at right angles. For example, for A = 3 meters, and B = 4 meters, at an angle of 90 degrees, the value of C = 5 meters. The early Greeks did not develop the cosine law for obtuse triangles. They solved the case of obtuse triangles by drawing a line from one of the angles to the other side at a right angle to the opposite side, thus cutting the obtuse triangle into two right triangles with a congruent side. The term or phrase 'right triangle' means any triangle with one angle equaling 90 degrees. The sum of all three angles of any plane triangle is equal to 180 degrees.

Very limited information is available on the early history of science much earlier than the time of Pythagoras.

A reasonable experiment to verify the Pythagoras formula is to draw a series of right triangles and measure the length of each side and compare the results from the equation $c^2 = a^2 + b^2$. A further experiment is to draw a series of obtuse triangles and verify the formula $c^2 = a^2 + b^2 - sqr(ab \cos c)$.

Figure 1 Triangle drawings. The reader is invited to enter his own drawings of various triangles.

Socrates

Socrates was primarily a philosopher and a promoter of the ideas of democracy in Athens. He held firmly to his teachings, but was condemned by a political group known as the 30 Tyrants. His death was from drinking poison hemlock made from a local plant (Conium maculatum, a member of the carrot family) .

Carrots, a root of the plant, are a healthy food source. Remove the plant tops. Many of the leafy parts of plants in the carrot family (Apiaceae) are not good to eat.

There are a number of toxic plants endemic or introduced around the world. Children should be taught about plants, their properties, which are good to eat and which are not. Children need to be taught which plants are toxic, and which are good to eat, raw or uncooked, and which must be cooked and for how long (such as potatoes). Information on which plants and which parts of plants are edible became well known from early times.

At the time of Socrates, Athens was a naval military power, developed in defense of the Greek islands. Sparta, located southwest of Athens across a bay, was ruled by an oligarchy, and had a strong army.

In Athens, good schools existed for boys (girls were schooled separately and trained to manage households). Sparta took over the teaching of boys at the age of seven, in army-like camps. Sparta stressed life in the army and did not teach mathematics or science to their youth. The Greek use of slaves captured in battles is beyond the scope of this book.

Reference: a good Wikipedia article on early Greek schools "Education in Ancient Greece"

Plato (428-347 BCE)

Much that is known about Socrates results from writings of Xenophon, and from Socrates' great student, Plato, who left much information in writing on papyrus scrolls. A problem in detailed history of that time period was that many records on papyrus paper were lost, such as in a fire that destroyed the Library of Alexandria, in Egypt. That fire was started by Julius Caesar during a naval engagement, and possibly was not intended to torch the Library.

Plato did not stress experiments, related to scientific observations. He was primarily a philosopher and allied himself with Socratic teachings.

Aristotle (384-322 BCE) See World Book Encyclopedia, Vol. 1, pages 663-664.

Aristotle was educated, for 20 years, in the school of Plato. After Plato's death in 347 B.C., Aristotle started his own school, the Lyceum, in 335 BC. Aristotle taught Alexander the Great, and was instrumental in the development of western philosophy. His ideas about an Earth-centered universe became a central idea in the development of early Christianity in the Catholic Church. Aristotle was primarily a philosopher, and did not stress experimental tests of his theories.

Refer to various encyclopedia, as well as articles on the internet for details of Aristotle's life.

Euclid of Alexandria (dates of birth and death are known only approximately).

Euclid was a great mathematician and developed a school in Alexandria. Euclid was a common name at the time of 450 BCE. Euclid organized a series of books concerning plane and 3-D geometry. These books stressed mathematical proofs of all theorems. The books were titled The Elements.

There were 13 books in the series of Elements. Some of the books were written partly by the best students of Euclid. The books were written on papyrus scrolls, and many copies were made for distribution. One of the most complete printing in English was in the translations by Sir Thomas Heath, a three volume set, and one of the Dover books on mathematics for advanced students. The Euclid books present a complete presentation of ideas on plane and spherical geometry and algebra. For discussions of mathematics covering conic sections, refer to later authors. The tables of contents for the three volumes are given below for Euclid's Elements. His Elements should not be confused with the chemical elements (refer to the periodic table of elements developed much later in the 18-th century).

Book One presents definitions, postulates and propositions related to the basic principles of points, lines, circles, parallels, triangles, rectangles and squares. The sections all begin with a numbered proposition followed by proofs. Proposition 1 gives a construction of an equilateral triangle, having three equal sides, given the base. Proposition 5 considers isosceles triangles which have only two equal sides. There are 48 propositions in Book 1. In the following, each of the books have Roman numbers.

A good way to study Euclid's Elements is to group the set of definitions, propositions and Lemmas by considering the plane geometry figures as inscribed in a circle (for example, a triangle in a circle). Similarly, but more complicated, for three dimensions, consider solid figures such as triangles, cubes, and multifaceted solids inscribed in a sphere (for example, a pyramid or a cube inscribed in a sphere).

Book Two continues where Book One left off. There are 14 propositions in Book 2.

Volume 2, Book Three deals with the properties of circles.

Book Four continues where Book Three left off, and developed concepts leading to extension of the earlier ideas of Pythagoras.

Book Five establishes proportions leading to commensurable and incommensurable magnitudes. Commensurable is a mathematical term one should refer to internet sources.

Book Six deals with applications and results of Book Five to plane geometry.

Book Seven is a complete introduction of number theory, and develops the Euclidean algorithm for finding the greatest common divisor of two numbers.

Book Eight There are 27 propositions in Book Eight

Book Nine There are 36 propositions in Book Nine.

Volume Three, Book Ten is a set of propositions dealing with commensurable measurements of points along a set of short lines. There are 115 propositions in Book Ten.

Book Eleven There are 39 propositions in Book eleven.

Book Twelve There are 18 propositions in Book Twelve.

Book Thirteen There are 18 propositions in Book Thirteen.

Some of the latter books may have been written by students after the death of Euclid, using Euclid's detailed outlines.

There are several so-called additional books, Books 14 and 15. See references by Sir Thomas Heath, pages 512 thru 519. The Heath set of three volumes include some of the original Greek language material. These volumes are complete and detailed. These three volumes may be used from beginning to graduate school studies of mathematics. Modern high school presentation of Euclid is given in books extracted from the three volume set documented by Sir Thomas Heath.

Apollonius of Perga (200 BCE) developed the mathematics of conic sections: the ellipse, parabola, hyperbola, and circle, leading to analytical and spherical geometry. Analytical geometry is an extension of simpler geometrical ideas by applying mathematical equations to the problems.

Spherical geometry deals with triangles on a sphere. Spherical triangles have three angles whose sum is equal to 180 degrees times 1+4f, where f is the fractional area ratio of the triangle to the spherical area of the sphere. Spherical geometry is important for navigation and for plate tectonics, as well as positions of stars and planets. Early astronomers had limited means to determine distances from Earth to the distant stars, using parallax measurements for nearest stars.

Eratosthenes (276-194 BCE)

Eratosthenes of Syene (near present day of Aswan) served as the third head librarian of the Library of Alexandria. He was also an astronomer. He measured (in about the year of 240 BCE) the circumference of the Earth using angles of the sun at Alexandria and Syene at noon at the summer solstice. Distances were measured by trained runners, or trained camels, and used times measured with a sun dial or water clock. Eratosthenes invented the concept of latitude and longitude. Ancient Syene is close to present-day Aswan and the Aswan High Dam.

Aristarchus of Samos (280 BCE) introduced the concept of a Sun-centered universe.

Hipparchos (190-120 BCE)

Hipparchos of Nicea was an astronomer who introduced plane and spherical trigonometry, trigonometric tables, and star maps for about 1000 stars, with Arabic names. He considered a Sun-centered Solar system. Hannibal was a great military leader, and is not dealt with in this book.

Hipparchos knew that several planetary orbits were not circular. He may have written 14 books, but none survived. Most of what is known about Hipparchos was based on Ptolemy's information.

Poseidonius (140-50 BCE) calculated the equatorial circumference of the Earth from erroneous measurements, and those values led Columbus to think Asia was about 3000 miles closer to Spain than in reality. Early sailing ships had to maneuver according to the wind, making distance measurements hard.

Early studies of medicine were made by Hippocrates and ideas of human behavior were summarized by Hippocrates, who used the four humors of sanguine, yellow bile, black bile, and phlegm, from the Arab world. Hippocrates wrote a early book on medicine, with help from his students. He was instrumental in describing the Hippocratic Oath used today. It stated that the medical treatment should do no harm. He new little about germs and viruses. He was able to set broken bones, but had no antiseptic or pain killers. His work was carried on by Galen up to AD 200.

Jesus Christ

Jesus Christ (born ca 0-10 AD) The followers of Jesus formed the beginnings of the Greek Orthodox (Eastern Orthodox) churches, and later the Roman or Catholic church.

100 BCE birth of Gaius Julius Caesar, who became a great military leader after Hannibal.

51 BCE Conquest of Gaul (southeastern Europe, present-day France to the Rhine River) According to the description of the area by Julius Caesar, Gaul is divided into three parts.

44 BCE Julius Caesar was assassinated to prevent him from becoming a dictator for life.

Ptolemy (90-168 AD)

Ptolemy of Alexandria (150 AD) published several books, called by others the Almagest documenting much of the known information on the subject of astronomy. The Almagest may have caused confusion about an Earth-centered versus a Sun-centered Solar system and the stellar universe.

Constantine (272-337 AD) His full name was Flavius Valerius Constantinus Augustus.

Mohammad (570-632 AD) develops the Islamic religion, and the Koran considered the true word of God given by the angel Gabriel to Mohammad between 610 and 632 AD.

The Bubonic Plague strikes Greece, Italy and southern Europe in 1328 and again in 1629 AD.

The 100-Year war between France and England caused the Roman Catholic Church to take drastic measures against heretics (1337-1453 and later).

Christopher Columbus (1492 AD) discovers the Americas by sailing west from Spain, landing in the area of the Caribbean. His plan was to find a fast way to reach India. He thought that the natives in the Americas spoke a dialect and had skin color lighter than many Eastern Indians. Thereby we still call the early natives in the Americas, Indians.

The Gutenberg Bible

Johannes Gutenberg (1395-1468 ? AD)

Gutenberg working with Johannes Fust and Peter Schoffer developing the first, practical printing machine using movable type, and published a printed bible in Mainz, Germany. His printing machine used wooden screws patterned after those used in the grape processing. There were several competing devices, at that time, but Gutenberg's press was the first to be operable. The Chinese and other groups had developed presses using carved wooden or ceramic character blocks, but too many such blocks were needed. Gutenberg's letters in his press were produced with the simpler German characters and created using the technology of precision metal working families in Mainz, noted for early expertise in jewelry manufacturing. (It should be mentioned that from the time period of mid-2000, most printing was done with lead characters cast from a metal form (reference articles on the Linotype machine).

Nicolas Copernicus (1543 AD) presented the concept of a Sun-centered universe, as an idea based on Aristarchus 280 BCE. No telescopic observations had been used for proof, before the time of Galileo and Kepler.

Giordano Bruno (1600 AD) was burned at the stake for his support of the Copernicus theory of a Sun-centered universe, a heresy at the time in the Catholic church.

Galileo Galilei (1564-1642)

Galileo is considered the first physicist, because he tested is ideas by experimental measurements, and was the first to use telescopic observations of stars, planets, our Moon, and our Sun.

Galileo Galilei (ca 1609) developed an improved version of a refracting (all lens) telescope, showing observations of Sun spots, a pocked-marked Moon, and made an observational proof that the universe was Sun-centered. He was attacked by the Catholic Church by Pope Urban VIII for heresy, and was placed under house arrest, and prohibited from all further publications. He had been given one of the telescopes to the Pope, with the idea "see for yourself". It took about 400 years for the Catholic Church to state that the Church had made a mistake, but that Galileo should have stated that the Copernicus Sun-centered idea was just a theory and not a true fact. Many of the churches at the time of Galileo claimed that their main documents were the incontrovertible word of God, and adhered to the teachings of Aristotle which held that the Earth was the center of the Universe. Aristarchus in 280 BCE and Copernicus in 1543 AD introduced the concept of a Sun-centered Solar system, as a theory.

Leonardo da Vinci (1452-1519)

Leonardo da Vinci is considered to be the first Renaissance Man. He was a great artist, an inventor ahead of his time, and a student of anatomy.

References:

Leonardo da Vinci, Flights of the Mind, a biography, 623 pages, by Charles Nicholl, 2004

Leonardo da Vinci, Renaissance Man, by Alessandro Vezzosi, 1997

Great Physicists, the Life and Times of Leading Physicists from Galileo to Hawking, William H. Cropper, Oxford University Press, Inc., 2001

Galileo and the Scientific Revolution, Laura Fermi and Gilberto Bernardini, Basic Books, INC, NY, 1961

Galileo, the Man and his Work, his Misfortunes, James Brodrick, S.J., Harpers and Row, 1964

Galileo: Watcher of the Skies, Wootton, David, Yale University Press, 2010

The early Catholic church held strictly to the concept of an Earth-centered universe, and considered the Sun-centered Solar system concept as a heresy. The Catholic Church, under Pope Urban VIII, held to the ideas of Aristotle of an Earth-centered universe, and prohibited the theory of Copernicus for a Sun-centered Solar system, considered by Galileo and Kepler (who fled to Prague to work under Tycho Brahe).

Martin Luther (1483-1546 AD) was a Catholic priest who successfully separated his new church from Catholic control.

Henry VIII created the Church of England, ending the rule of the Roman Catholic Church in England (1531 AD).

Johann Bayer (1572-1625)

Prior to the time of Galileo, all astronomical measurements were carried out without the aid of a telescope. Bayer published an early catalog of stars and other bright objects in the sky, by 1603. Bayer was born in Swabia, Germany and died in Augsburg. Bayer measured about 1600 of the brightest stars and established the identification of visual magnitudes in order of the brightest in each constellation with a lower-case Greek letter, starting from alpha for the brightest star.

Galileo obtained an early telescope developed by a North European lens maker named Hans Lipperschey, and used it to make carful astronomical measurements of the stars, planets, and the Moon.

Johannes Kepler (ca 1610) makes further improvements of Galileo's refracting-lens telescope, and makes a strong case against an Earth-centered universe. He managed to escape from Italy to northern Europe, where Martin Luther as a German Catholic priest had separated his Lutheran Church from the Roman Catholic Church control.

Johannes Kepler

Kepler published a correct theory of tides which resulted from the attractive force of gravity due to the orbital motion of the Moon. It may be mentioned that Galileo argued (incorrectly) that winds were the sole causes of the tides, and that the orbital motion of the Moon was not responsible for the tides.

Kepler established a set of three laws governing the planets. Using telescopic and visual measurements, based partly on measurements obtained by Tycho Brahe, he proved that the planets of the Solar system moved around the Sun in elliptical orbits, with the Sun at one of the foci of the ellipse. The second law states that a line from the planet to the Sun sweeps out equal areas in equal time. The third law states that the square of the planetary period is proportional to the cube of the semi-major axis of the orbit.

Isaac Newton (1632-1727 AD) developed the laws of motion, gravitation and the theory of light using the double prism proving that the colors of sunlight passed through a prism are not further changed by the second prism.

Newton's laws are as follows: Force F=mass times acceleration; Kinetic energy = mass times velocity-squared divided by 2; and the force of gravitation = G times m1 times m2 divided by separation, r-squared. He defined the constant of gravitation G.

He developed a reflection telescope with an ocular on the side, 90 degrees from the main optical axis. The Newtonian reflecting telescope using mirrors instead of lenses, reduced the chromatic aberration caused by the primary optical material. Newton's design, however, did use glass lenses in the eye piece, which introduced some off-axis distortion of the images beyond about one degree from the center of the field of view.

Newton's telescope used a parabolic, concave metal primary mirror, and a smaller convex metal mirror, plus an eye piece or ocular glass lens. The reflector design removed chromatic aberrations, but introduced off-axis spatial aberrations resulting in a narrow field of view for angles greater than about 1 to 2 degrees. This aberration is called commatic aberration, because stellar images were not imaged as points, but had a comma-like tail.

A change in reflection telescope design was introduced by Gregory in the decades after Newton, which used parabolic mirrors. The Newtonian reflector used a concave parabolic primary and a convex secondary mirror. The Gregory reflection telescope used the concave parabolic primary mirror, but a concave secondary mirror. Modern versions of the reflection telescopes are referred to as a Cassegrain telescope, and are discussed in later sections of this book.

Models of the Galilean refraction telescope and cutaway models of the reflecting telescopes are available from the internet and in many science museums, such as the Griffith Observatory in Los Angeles.

A later version of reflection telescope designed by Cassegrain used a hyperbolic secondary mirror. In the 1920 time period, another version of the reflection telescope was introduced by Ritchey and Cretien using hyperbolic primary and secondary mirrors, which introduced complications in the grinding and polishing of the mirrors. The Mt. Wilson 60, and 100 inch as well as the Mt. Palomar 200 inch telescope avoided that design which required new ways to grind and polish the mirrors. The original Hubble telescope which used the Ritchey-Cretien hyperbolic mirrors were incorrectly polished and resulted in poor performance which required astronaut correction in an extra-vehicular space walk to install a complex system of lenses designed by JPL.

The details of the corrected optical system for the Hubble telescope should be discussed in a later version of this book.

For the history of American War of Independence (1775-1783), is beyond the scope of this book, and is covered in Wikipedia articles, and many history books.

The Age of Enlightenment beginning with Martin Luther, Kepler and Newton are covered in chapter 3.

Vectors

A vector is an expression of a magnitude plus a direction. There are two kinds of vector products: the scalor or dot product A . B = A(magnitude) times B(magnitude) times the sine of the angle between. And the vector or cross product: A x B times cosine of the angle perpendicular to the vertors A and B. See reference on Wikipedia and Hyper physics.

THE AGE OF ENLIGHTENMENT

This chapter covers the rise of the ideas of advanced aerodynamic flight, the electron field theory of Maxwell, and the rise of quantum mechanics starting with Michelson and Morley, Max Planck, Albert Einstein, Niels Bohr, Schrodinger, Heisenberg, Dirac, Pauli, Fermi, Hermann Hess who discovered cosmic rays, Curie who isolated radium and polonium, Irene Joliot-Curie who discovered induced radioactive species, and Willard Libby and others who found methods of dating rocks, fossils, and other animal and plant remains younger than about 40,000 years, and with isotopic methods to date older rocks; high energy particle accelerators (cyclotrons, Van de Graaf accelerators, linear accelerators, Tevatrons, CERN; and extensive tabulations of radioactive emissions which led to the development of modern physics of sub-atomic particles. The development of electronics and radio by Marconi, Armstrong, and de Forest, became a troublesome patent lawsuit problem up to the time of the 1929 depression.

Modern physics developed during the middle to late 1800's in the attempts to explain the spectral emissions and lines of high-voltage excited gasses, and the light emitted by hot gasses in an oven. The light from stars was found to include spectral lines as well as a broad spectrum of light, known as thermal blackbody radiation. Sound is propagated in air as a wave, and the theory of waves in air or water was well understood. But how light travels through the vacuum of space from stars was not understood. There was no media for light in a vacuum. If light were a wave, there must be an unknown wave medium. Until a medium is discovered, one was invented, and called the luminiferes aether. A search for the aether was undertaken by measuring the variation of light speed with direction and time. In short, no such variations were found. Light had both wave-like properties as well as particle-like properties.

Albert Michelson and Edward Morley conducted a long set of precise measurements by 1887 to discover the variations of the speed of light with direction of propagation, or time. No such changes were found. The speed of light was a constant. The aether was found not to exist. The era of modern physics had begun.

The next developments included the equation for blackbody radiation by Max Planck, and quantum science by Albert Einstein (the special theory of radiation in 1905 and the general theory of radiation in 1915). Spectral emissions and line spectra were determined by the Rydberg, and other formulas.

Modern physics is easier to understand conceptually than to be explained mathematically, which is usually deferred to graduate school, due mainly to the lack of advanced calculus and differential equations (which can be introduced as early as high school). The modern approach to teaching science is to introduce the science without the math; then, years later introduce the ideas of calculus and differential equations and matrix theory. In today's internet age, teaching of mathematics may be introduced in grade school, and in high school, and not wait to the fifth year in college (graduate school).

Quantum Mechanics

The first problem in the introduction of quantum mechanics is that the main concepts, although simple are not intuitive. The problems include the ideas of step-wise, discontinuous energy transitions and probability amplitudes, and spin. Another problem is the lack of familiarity with the mathematics, including differential equations, usually introduced in graduate school (way too late for good students).

Modern Astronomy

Astronomy has led to important discoveries over the past 3000 years in the fields of science. The Mt. Wilson and Palomar telescopes played a great part in development of current understanding of the universe. Using the telescopes on Mt. Wilson, stellar distances were determined. In the most recent 20 years, a number of large telescopes have been or are being built, leading up to a 30-meter diameter TMT telescope being built by the University of California, with first light planned for 2018 to 2021, if budgets survive. Larger aperture telescopes are being planned by a European astronomy group (with a 43-meter multi-mirror primary).

Spaced-based astronomical telescopes became available with the Space Shuttle-launched Hubble telescope in low-altitude Earth orbit, and later beginning in 2005, deep-space telescopes—the Spitzer, Hershel, and several more recent telescopes, leading to the future James Webb telescope, all with infrared and ultraviolet capability, not possible within the atmosphere of Earth.

A highly important advance in astronomy is the recent development of space-based telescopes which include the Shuttle-launched and serviced Hubble telescope presently functioning in low Earth Orbit. The Hubble was followed by the Spitzer, Hershel, Kepler, and the future James Webb deep-space telescopes. Solar orbiting satellites were launched capable of monitoring sunspots and flares on both sides of the Sun, at the same time.

Reference: www.solarham.com

Of considerable importance since 1993 is the development of automated, remotely controlled, and internet-connected 0.4 to 2 meter telescope systems available to the public and educational groups throughout the world. Astronomers no longer have to fly around the world, and wait for night and clear weather. They can do some astronomy from the office or school classrooms, using an automated telescope operating thousands of miles away where it is still dark, and skies are clear.

The early twentieth century was also the time when many modern science developments as well as the creation of manned flight aircraft, advanced radio transmitters and receivers, high-fidelity radios, television and greatly improved nuclear radiation detection systems, modern electronics based on semiconductors, rockets, orbital spacecraft, geological plate tectonics, nuclear reactors, nuclear weapons, the modern automobiles, radar, large aperture telescopes which lead to the discovery of an expanding universe, modern home appliances, and the discovery of DNA, and the Van Allen Belts of geomagnetically trapped corpuscular radiation (electrons and protons), and solar corpuscular radiations (1 to 100 MEV protons), muons, and the Solar wind (low energy protons and electrons).

Solarham.com

LCOGT.com Las Cumbres Observatory systems

Hess, Wilmot, Trapped Corpuscular Radiations

The Michelson-Morley Experiment (ca. 1887+)

The concept of the wave-particle duality of light had not been developed fully, by 1900. If light was a wave, the wave behavior required a medium. No medium was known, so one was invented-the aether. If aether exists, it should be detected by precise light velocity measurements in different directions and measurements over the course of a year. After many careful attempts, no variations in the speed of light were detected, even with respect to direction of propagation. Rather than to eliminate the wave theory of light, attempts to explain the behavior of light were made, and finally it was decided that light has both wave and particle explanations.

Reference: The Michelson and Morley, the experimental search for an luminiferous aether.

The Development of Quantum Mechanics Mathematical Expressions of Heisenberg, Schrodinger and others.

Max Planck and the Black Body Formulas

Late 19-century scientists struggled with limited success to characterize black body radiation theory. Planck solved the problems by showing that the black body radiation resulted from quantized sub-atomic material oscillators. Rayleigh-Jeans produced good analyses for low frequencies and Wien produced analyses for high frequencies. Planck solved the problem, and in the process, showed how quantization of the material oscillators (primarily electrons) was required. What he showed was that the sub-atomic processes of energy proceeded in quantum jumps, not in the smooth transitions of macroscopic energy of more familiar energy transitions.

In 1901 when Planck first published his blackbody formula, the concept of quantization of light and of material oscillators had not been fully developed. Albert Einstein and others developed a theoretical process leading to Planck's formula based on first principles (quantum theory). The concept of the wave-particle duality of light was introduced. How could light be both a wave without a medium, or a particle?

Light refracted by lenses, or in air can be considered as a wave. Light in a vacuum and observed on the small scale in an experiment, can be considered as a particle. Radio wave transmission to Mars and other planets exist in the form of particles (photons). Radio signals are viewed primarily as waves.

Ernest Rutherford

Rutherford discovered upon irradiation of a thin sheet of gold with alpha particles, that the atoms of gold consisted of tiny, sub-atomic particles and lots of space in between.

References:

Rutherford, Ernest, the Gold Foil Alpha Scattering Experiment, 1911

Radio Receivers and Transmitters

The science of radio began at the turn of the century between 1887 to 1933 with the development of vacuum tubes and electronics. The detailed history of radio, electronics and television is beyond the scope of this book. One should refer to the many books and encyclopedias for the details.

Vacuum Tube Development

Radio Transmission and Receiver Development

Television Development

Large Scale Integrated Circuit Electronics

The development of large-scale integrated circuits had a major effect on electronics and mathematical science, and allowed the IBM and other companies to provide important computers. In the 1950's IBM developed the punch card machines. Punch cards could be created by using special pencils to mark the cards as well as using a machine to put small rectangular holes in the cards which were used for running big IBM and other high-speed computers. It became a standard for a scientist or businessman to fill out what was called a green-sheet, which allowed a key-punch operator to provide a set of cards for driving large, main-frame computers that were very fast and efficient. These high-speed machines revolutionized mathematics using numerical methods to solve complex mathematical calculations not previously possible. Later IBM, Bill Gates, and others established the first efficient personal computers.

The history of these new computing machines had a major effect on business and science.

Such machines revolutionized physics, and engineering.

Millikan, Robert, Andrews, Duane Roller, and Ernest Watson, Elementary Physics, 1937

Electron charge and mass determination by Millikan's oil drop measurements

There were major advancements in nuclear physics and chemistry with advanced discoveries by Planck, Einstein, the Curie family, Chadwick, Anderson, Shroedinger, and others in the 1900 to 1940 time period.

Introduction to Modern Physics, Richtmeyer, F. K. and E. H. Kennard

Introductory Nuclear Physics, Halliday, David, 1950

The Feynman Lectures on Physics, three volumes, Richard Feynman, 1960

The Theory of Quarks, Gell-Mann, Murray, various papers, and Nobel lecture, 1964

Gell-Mann, Murray, the Quark and the Jaguar,

This book is an excellent introduction to modern physics.

Hawking, Stephen, W., A Brief History of Time, 1988

Hawking, Stephen, W., The Universe in a Nutshell, 2001

Arrhenius, Svante, Computations on Climate Change, varius publications, 1899

Bohr, Niels, The Hydrogen Atom,

Planck, Max, (1858-1947 AD) the Black Body Formula, 1901, p 67

Reference the internet for Max Planck.

Einstein, Albert, The Photoelectric Effect,

Einstein, Albert, Special Relativity Theory, 1905

Einstein, Albert, General Relativity Theory, 1915

Heisenberg, Werner, The Physical Principles of Quantum Theory, translated by Carl Eckart and F. C. Hoyt, Univ. Chicago, and Dover Publications, 1930

J. J. Thompson (1856-1940) Nobel lecture 1906

Thompson discovered the electron in 1897. His atomic theory is known as the plum pudding model.

Niels Bohr (1885-1962), The Bohr Hydrogen Atom, Nobel Lecture, 1922

Great Physicists, by William H. Cropper, Oxford University Press, New York, 2001

The first attempt to describe the hydrogen atom used an analogy to planetary motion, where an electron orbits around a small, positively charged, nucleus of protons and neutrons. A main problem solved was that in classical theory, an orbiting electron would slowly loose energy to x-rays in the process of bremsstrahlung (freely translated as electron braking-radiation). This was the difficulty with the earlier Rutherford model of the atom.

The new idea was that the electron could emit energy only in discreet steps (called quantum steps, without emitting a continuous stream of bremsstrahlung x-rays). The early Bohr theory for hydrogen required further development for heavier (higher atomic number) elements.

Further improvements in the model of atoms needed explanation of the differences between atomic number and atomic mass. Much of the answers depended on the discovery of the neutron by Chadwick in 1932.

Chadwick (1891-1974)

Rutherford suggested that elements heavier than hydrogen had nuclei containing positively charged protons plus a number of unknown and uncharged particles which accounted for the total atomic mass of elements greater than hydrogen. At that time, all radiation detectors worked only for charged particles or x-rays. There was no known detector for these unknown, uncharged particles, until the mid-1900's.

Ernest Rutherford suggested the possible existence of an unknown, uncharged particle—the neutron, in 1920.

In 1931 Walther Bothe and Herbert Becker in Germany found that the alpha particles emitted by polonium were used to irradiate beryllium, boron, and lithium, that an unidentified high energy radiation was produced. It was originally thought that this radiation was an intense gamma ray.

In 1932, Irene Joliot-Curie and Frederick Joliot in Paris also noted the unknown radiation produced when alpha radiation was used to irradiate aluminum, and that the radiation produced was greatly increased when a block of paraphine was placed on the aluminum. This led to the suspicion that a new particle, such a Rutherford's neutron was being created, and the neutron was a part of many nuclei.

Later in 1932, James Chadwick carried out a series of experiments at the University of Cambridge, proving the new radiation had nearly the same mass as the proton, but had no charge. He identified the new particle as the neutron.

Chadwick receiver a Nobel Prize in physics for the discovery of the neutron

It was subsequently discovered that the free neutron would undergo a spontaneous disintegration via a beta decay with a half-life of about 12 minutes.

Robert Andrews Millikan (1868-1953)

Millikan conducted a long-duration effort from 1908 to measure the charge and mass of the electron, assuming that the electron was a particle and not a wave. The measurements continued with improvements, over a period of years, using a troublesome oil drop measurement method. The best setup used ultra-low viscosity watchmaker's oil. The oil was dropped into a chamber containing a pair of electrical plates charged to a high voltage, adjusted so that a single drop could be held between the plates by electrostatic attraction after receiving an electron charge. The charge varied from a single electron to a number of electrons. Statistical procedures were used to determine the mass and charge for an individual electron. The results were subject to some controversy from other researchers who felt that there might exist sub-electrons, never found in Millikan's measurements.

Reference:

Brian, Denis, Albert Einstein: a Life

Einstein's Luck: the Truth Behind Some of the Most Important Scientific Discoveries, by John Waller

This book reads like a critique of scientific data handling problems, but in fact it is well written and documented, but it could present more of the mathematics. It deals correctly in reference to Millikan's constant difficulties, as well as those of Gregor Mendell, Charles Darwin and a number of other 20-th century scientists.

Reference: Thomson, G. P. Thomson and J. J. Thomson, Discoverer of the Electron, 1897

J. J. Thompson (1856-1940)

Thompson discovered the electron in 1897. His atomic theory is known as the plum pudding model.

This theory was overturned by Rutherford, Niels Bohr and Einstein.

James Chadwick (1891-1974)

Chadwick identified the neutron as another part of elements greater than hydrogen.

Anderson, Carl David (1905-1961)

Anderson identified the positron in radioactive decay.

Irene Curie (1904-1956)

Irene, daughter of Marie, received a Nobel Prize in chemistry for the discovery of induced or artificial radioactivity. She used an alpha radiation source to impact a thin aluminum foil, which resulted in the generation of neutrons, which in turn irradiated a target of a different material, resulting in a process of creating a new, radioactive element, as evidenced by the exponentially decreasing count rate of a Geiger counter, which continued after removal of the alpha and source of neutrons.

References:

Curie, Eve, Madame Curie, a biography, translated by Vincent Sheean, 1937

Brian, Denis, The Curies, a biography of the most controversial family in science, 2005

Albert Einstein developed both the general and special theories of matter, light, and quantum theory.

Fermi-Dirac Statistics

Nuclear Particle Detectors and Electronic Amplifiers

A good discussion of radiation detectors, amplifier circuits (vacuum tube technology) and radiation sources is given in Halliday. Detailed circuits with vacuum tube technology are given in the book by Mathew Sands, and others of Caltech.

In December 1947, Bell Labs published that they had perfected the first solid state electronics amplifier and circuits, and had constructed a small radio using semiconductor circuitry by spring 1949.

During World War II, essentially all electronics was via vacuum tubes.

Reference:

Sands, Mathew, et al

The first particle detectors were the fluorescent screen, the ion chamber detector, the Wilson cloud chamber, and photographic emulsions. Later the ion chamber detector, proportional counters, and Geiger-Mueller counters, were developed. In the 1960's solid state detectors were developed.

By the 1955 time, solid state electronics began to replace vacuum tube technology.

Hess, Victor (1883-1964)

Victor Francis Hess carried ion chamber detectors on balloon flights up to 17,000 feet altitude.

He wanted to measure the decrease in radiation ion rate with altitude. What he found was that the ion rate increased with altitude, due to an unknown source of radiation beyond the Earth's atmosphere. He called that source, cosmic rays. The years of repeated flights on manned balloons up to altitudes of about 18000 feet, was in 1911 to 1912. He received a Nobel Prize for his discoveries.

Following that, Millikan, Neher and Anderson and others launched unmanned balloon flights every summer, using an automated ion chamber and an early form of radio telemetry. Flights were made from stations near the north and south magnetic poles.

The ion chamber measures a rate of ionization from radiation, but does not give actual count rates or particles per second, and does not indicate the type of energy or type of particle.

Later in the mid-1950's, Van Allen flew a pair of Geiger counters on rockets up to altitudes of 10,000 miles. Geiger counters can count individual particles, but can't determine the particle type. To determine particle type, one Geiger counter had a thin wall making it sensitive to electrons and protons, and the other had a thick wall, making it sensitive mainly to high energy protons. The Geiger counter detectors were also sensitive to x-rays and cosmic rays. What he discovered were intense regions of protons near an equatorial altitude of 2000 miles, and intense regions of electrons above an equatorial altitude of 10,000 miles. These regions of radiation were called trapped particle belts, because they were located in narrow zones along the Earth's magnetic field lines. Trapped protons and electrons are locked into spiral orbits around the Earth's magnetic fields, and can move (that is trapped along the Earth's magnetic field lines) along the field lines to the north and south polar regions.

Later measurements were made with cloud chambers, scintillation detectors and solid-state detectors which confirmed the particle types and energy levels. Detectors were flown on orbiters as well as rockets.

Black Holes, Dark Matter, and Dark Energy

The theory of Black Holes was developed by Hawking and others at the California Institute of Technology.

Hawking, Stephen W., a Brief History of Time, Bantam Books, 1992

This book was on the best seller list for 4 years.

Theodar Von Karman (1881-1963)

Karman developed the theory of the interaction between the air and the surfaces of aircraft.

Ludwig Prandtl and Karman developed the theory of high speed flight aerodynamics.

Liepmann, H.W. and Anatol Roshko:

Liepmann, H. W. and Anatol Roshko, Elements of Gasdynamics, Courier Dover Publications, 1993

SPACE MISSIONS

Large Rocket Weapons

Near the end of World War II, Dr. Werner Von Braun in Germany, developed the concepts of large military rockets. The first was the V1 rocket, a modified, unmanned aircraft with a small rocket motor, followed later by the V2 rocket. Both carried high explosive payloads. They were used against the southern parts of England. The V1 engine was noisy. It was called the buzz-bomb, and it gave a short warning followed by silence as the rocket motor stopped, prior to crashing and exploding. The V2 rocket launched in a high altitude trajectory, and gave no warning sound prior to hitting the target and exploding. By the time radar had detected it, it was too late for anyone to take cover.

The V-1 and V-2 were among the first liquid-fueled large rockets. Dr. Robert Goddard developed the early rocket motor technology. The letter V was from the German words, "Verteidigunds waffe" translated to English, "Defense Weapon" or "Revenge Weapon".

Rocket Development in the United States and in Russia

Within about 10 years after the second world war, rocket experiments were being developed by the US and Russia with inter-continental range. The first space system was the Russian orbiting satellites--the Sputniks, in October 1958. These first Earth orbiting satellites came as a shock to many Americans who felt the US was losing the space race, and many feared that future satellites might secretly contain nuclear weapons.

Within the next 6 months several US rockets carried Geiger counters designed for detection of high energy radiations. These detectors were developed by Van Allen of the State University of Iowa. These Geiger counters discovered zones of magnetically trapped high energy protons and electrons. The 100-MEV-range of protons were detected at an altitude of about 2000 miles above the equator. The 1-MEV-range of electrons were at an altitude of about 10,000 miles above the equator. These particles circle around the geomagnetic field lines, spiraling to the north and south, down to the polar regions, then going back up. Later satellites discovered high energy protons were frequently directed to the Earth following major flares on the Sun. These radiations posed unknown hazards for spacecraft electronics and future astronaut flights. Fortunately it is possible to shield against such radiations, mainly by requiring a minimal thickness of the command module of 3 grams per square centimeter, aluminum plus heat shield materials. Shielding against high energy protons can be accomplished by use of water-filled plastic blankets with an aerial density of at least 10 grams per square centimeter (about 4 inches in thickness).

The first major manned spacecraft program was established by President John F. Kennedy, the Apollo Lunar landing and return missions. The Apollo missions were to be scientific missions of discovery and a drive towards development of large rocket launch vehicles. The Apollo project was established as a scientific and engineering breakthrough, development effort. The emphasis was on high reliability. It proceeded with minimal accidents. The worst accident was the fire during a ground-based test of the command module in January 1967. The fire was as a result of the use of 100 percent oxygen in the command module, an inherently hazardous procedure that resulted in the death of three astronauts, Gus Grissom, Chaffee, and White. I recalled no specifications for design of the command module stipulating that there would be a 100 percent oxygen test, although the concept for use of increased oxygen on the launch pad was considered. The discussion of the tragic launch pad test fire is beyond the scope of this book. The problems of the transition from normal Earth to low oxygen at high altitude to space had been solved. The astronauts could live on an atmosphere of 100-percent oxygen at about 3 percent of normal, sea-level pressure.

Solar Corpuscular Radiations

Shortly before the initial design stages of the Lunar Mission with Apollo, high energy Solar protons were detected in space following certain Solar Flares. Apollo missions operated during a period of low Solar proton events, and the astronauts needed no radiation shielding to protect them. Had strong proton events been expected during space flight, plans included water blankets filled with 5 to 10 centimeters of water to cover the full body of astronauts for protection during a Solar proton event.

These water blankets would have been helpful during traversal of the trapped radiation zones, but at the high velocity of that traversal, the dose to the astronauts was small. The water blankets were not needed for the short duration of the Apollo missions during periods of low Solar activity.

The Apollo Lunar missions were free of accidents. The only accident was on a launch pad test, prior to the Lunar missions. The discussion of the launch pad test fire is beyond the scope of this book.

Apollo 13 had a internal explosion of an oxygen tank in the service module. The Lunar landing by Apollo 13 was prevented. All astronauts returned to Earth safely.

The DNA Molecule

The chemical structure of the fundamental substance behind all living mammalian and plant cells was determined by x-ray diffraction spectroscopy by Rosalind Franklin in the early 1950's. Her work was carried on by Watson and Crick, and published after Franklin's death. A Nobel prize in physiology was awarded for this ground-breaking discovery of the molecular structure of the DNA molecule.

James D. Watson and Francis Crick, The Double Helix, and the Secret of Life, 1998

The DNA structure can be visualized as a twisted, sub-microscopic stepladder. It is too small for many electron microscopes, which tend to break up the DNA structures. Dr. Rosalind Franklin established the structure using a highly-advanced, x-ray diffraction technique. Franklin established the DNA structure as a twisted, double helix. The basic DNA molecule consists of two sugar-phosphate structures with pairs of cross-structures composed of molecules of adenine, cytosine, thiamine, and guanine, taken two at a time.

The Space Shuttle

The concept of the Space Shuttle was to provide a launch capability for deploying scientific equipment and astronauts with aircraft-launch rather than the costly, large rocket launch vehicles. Large rockets are costly, and single-use systems. To reduce costs, the shuttle was to be launched by a large aircraft, which would return to Earth, landing on an airfield. The Shuttle would have small, rocket engines to carry it to higher altitudes. The Shuttle was to return to Earth and land, and could be reused.

Development-cost problems were encountered. A larger launch aircraft had to be designed, built and tested.

The largest available jet engine aircraft, the Boeing 747, without passengers, was not capable of carrying the Shuttle, with crew and equipment. Although the 747 could carry the Shuttle without crew, fuel, and equipment, the 747 was used to carry the Shuttle back from Edwards Air Base to Florida for the next Shuttle flight. The 747 could not carry the Shuttle with astronauts, fuel, and equipment. Larger launch rockets would have to be used.

The alternative Shuttle Program concepts included a new Shuttle, one for equipment and another for the astronauts and/or a single aircraft capable of launch from an airfield directly to orbit. Not all space satellites would require the Astronaut for operation in space. Not all satellites would use the same altitude, and would need the astronauts. Such systems are in development.

The Hubble Telescope was launched into a low-altitude orbit, and was serviced by astronauts, on later flights. The very successful Hubble was the last large telescope put into orbit and serviced by astronauts.

Other space telescopes were launched by large launch rockets, and put into distant orbits. This meant that when the telescope system needed a new set of reaction gyros, coolant fuel resupply, or repair, it was the end of the telescope. The future large Web telescope for example is to be placed into a solar orbit beyond the reach by astronauts launched by the Shuttle. Long term system reliability would have to be achieved only by a replacement, or a robotic repair system, not presently being considered.

Another example is the GPS spacecraft (NAVSTAR) satellites in a high altitude orbit. Originally, the Shuttle was to release the NAVSTAR into orbit. It was decided that the NAVSTAR would have to be launched by a large, unmanned launch rocket. That meant that the original orbital altitude and inclination had to be changed, and the Shuttle/aircraft launch could not be used. The main use of the Shuttle was limited to delivery of equipment and astronauts to the manned, International Space Station.

SPACE MISSIONS

Large Rocket Space Vehicle Systems

Near the end of World War II, Dr. Werner Von Braun, in Germany, developed the concepts of large military rockets. The first was the V1 rocket, a modified, unmanned aircraft with a small, pulsed-jet rocket motor, followed later by the V2 rocket. Both carried high explosive payloads. They were used against the southern parts of England. The V1 engine was noisy. It was called the buzz-bomb, and it gave a short warning followed by silence as the rocket motor stopped, prior to crashing and exploding. The V2 rocket launched in a high altitude trajectory, and gave no warning sound prior to hitting the target and exploding. By the time radar had detected it, it was too late for one to take cover.

The V1 rocket, or the Buzz-bomb, could be called, in today's language, the first drone aircraft bomb used in a war. It was tested in development by a woman pilot who survived the test flight, and the war, and, later, toured the United States.

The V-1 and V-2 were among the first liquid-fueled large rockets. Dr. Robert Goddard developed the early, experimental, non-solid, rocket motor technology. The letter V was from the German words, "Verteidiguns Waffe" translated to English, "Defense Weapon" or "Revenge Weapon".

Rocket Development in the United States and in Russia

Within about 10 years after world war-2, rocket experiments were being developed and tested by the US and Russia with inter-continental range. The first orbiting spacecraft systems were the Russian orbiting satellites-- the Sputniks, in October 1958. These first Earth orbiting satellites came as a shock to many Americans who felt the US was losing the space race, and many feared that future satellites might secretly contain nuclear weapons. As it appears, the cost of producing and testing of a high priced, secretive, large, nuclear weapon, that must be defended in space-orbit, was not going to compete favorably with Inter-continental Ballistic Missiles, the ICBM's.

The question I had taken on early in my career in aerospace was what mix of military aircraft and rockets would be needed for future international military needs. ICBM's, high speed aircraft, and radar-eluding aircraft must be included in the mix of defense weapons. ICBM's, after launch can't be stopped. They can be controlled after launch not to explode, but they can't be recalled to return to base. High speed, Mach 1, or greater, aircraft can be recalled to return to base, and can avoid radar, but not infrared seeking missiles, due to the high aircraft surface temperatures associated with high speed. Submarine-launched and carrier-launched missiles and aircraft are needed for pre-war weapon storage, world-wide. The details of such considerations are highly classified and far beyond the scope of this history book.

Within the mid-1950's several US rockets carrying Geiger counters designed for detection of high energy radiations were flown. The first radiation detectors in space were developed and flown on small rockets by Van Allen of the State University of Iowa. The Geiger counters discovered zones of magnetically trapped high energy protons and electrons. The 100-MEV-range of protons was detected at an altitude of about 2000 miles above the equator. The 1-MEV-range of electrons was at an altitude of about 11,000 miles above the equator. These particles circle around the geomagnetic field lines, spiraling to the north and south, down to the polar regions, then back up again. Later satellites discovered that high energy Solar protons were frequently directed to the Earth following major flares on the Sun. These radiations posed unknown hazards for spacecraft electronics and future astronaut flights, and the discovery came as a shock to all scientists. Fortunately it is possible to shield against such radiations, mainly by requiring a minimal thickness of the command module of 3 grams per square centimeter, aluminum plus heat shield materials, leaving the bremstrahlung x-rays. Shielding against high energy protons can be accomplished by use of water-filled plastic blankets of at least 10 grams per square centimeter (about 4 inches in thickness, leaving the secondary neutrons). This shielding would be needed only during traversal of the radiation belts, and for a few hours during and after certain large solar flares, forming on positions of the Sun directed towards the Earth.

Prior to the Apollo missions, low altitude, short-duration manned flights were undertaken as a test of the ability of man to go into space. The problems of the trapped electron and proton radiation zones needed much more measurement by orbiting satellites before any manned missions were launched.

The first major manned spacecraft program was established by President John F. Kennedy, the Apollo Lunar landing and Return Missions. The Apollo Missions were to be scientific missions of discovery and a drive towards development of large rocket launch vehicles. The Apollo project was established as a scientific and engineering breakthrough-development effort. The emphasis was on high reliability. It proceeded with minimal accidents.

The only fatal accident was the fire during a ground-based test of the command module in January 1967. The fire was as a result of the use of nearly 100 percent oxygen in the command module, an inherently hazardous procedure that resulted in a flash-fire and the death of three astronauts, Grissom, Chaffee, and White. I knew of no specifications for design of the command module stipulating that there would be a 100 percent oxygen test on the launch pad with a crew, although the concept for use of increased oxygen on the launch pad had been considered as a way to optimize the amount of oxygen needed during launch and flight. The discussion of the tragic launch pad test fire is beyond the scope of this book (for example, why was there a full crew of astronauts for a simple launch pad test--with nearly 100 percent oxygen in the command module, a test that could have been conducted without a crew)?

Solar Corpuscular Radiation Events

In the event of a major Solar flare, there would be ample warning from ground-based solar telescopes. No Solar flares were seen immediately prior to Apollo launch times which were during a period of low solar activity. Large proton events that would impact the Earth, were un-predictable (only certain such proton emissions that could reach the Earth, had to have been from large sun-spot groups located in a narrow region around the Sun from which the protons would be directed towards the Earth. Most proton radiations are deflected into space beyond the Earth. More than half of such events, on the other side of the Sun never reach the Earth. The geometry indicates that there is at least a fifty-percent safe window in time, for danger from a single, major event. During the maxima in Solar activity, there might be several large sunspot groups visible from the Earth, at any one time.

Shortly before the initial design stages of the Lunar Mission with Apollo, high energy solar protons were detected in space following certain large solar flares. But the Apollo missions operated during periods of low solar flare events, and the astronauts needed no radiation shielding against small Solar or trapped protons. Had strong proton events been expected, the launch could have been delayed, or the use of water blankets filled with 10 centimeters of water to cover the full body of astronauts for protection during a major solar proton event, which was unexpected during a time of minimal solar activity.

These water blankets might have been helpful during traversal of the trapped radiation zones, but at the high velocity of that traversal, the expected radiation dose to the astronauts was small. The water blankets were not needed for the short duration of the Apollo missions during periods of low solar activity. Sufficient shielding was due to the amount of materials making up the walls and heat shielding of the command module, which amounted to somewhat more than 3 grams per square centimeter in thickness.

The Apollo Lunar missions were free of fatal accidents. The only fatalities were on the launch pad test, earlier than the Lunar missions. The discussion of the launch pad fire and the justifications for such a test conducted with astronauts in the command nodule is beyond the scope of this book.

Apollo 13 had a minor, internal explosion in an oxygen tank in the service module. The Lunar landing by Apollo 13 was aborted and all three astronauts returned to Earth safely.

The International Space Station Created with Astronauts as a Scientific follow-on to Apollo

The ISS was constructed from modular components launched by the Space Shuttle. The Space Shuttle carried both equipment and a crew of three. The ISS was launched into a low-altitude (200-300 mile altitude) in a near-circular orbit with an orbital inclination of about 32 degrees.

References

Oberg, James, The International Space Station, World Book Online, 2007

Nelson, Maria, Life on the International Space Station. 2013

Apollo Missions:

Apollo 7 1968 Oct 11

Apollo 8 1968 Dec 21

Apollo 9 1969 May

Apollo 10 1969 May 18

Apollo 11 1969 July 16 First Lunar Landing July 20 Armstrong, Collins, Aldrin

Apollo 12 1969 Nov 14 Conrad, Gordon, Bean

Apollo 13 1970 Apr 11 Oxy tank event No Landing by Lovell, Swigert, Haise

Apollo 14 1971 Jan 31 Shepherd, Roosa, Mitchell

Apollo 15 1971 July 26 Scott, Worden, Irwin

Apollo 16 1972 Apr 16 Young, Mattingly, Duke

Apollo 17 1972 Dec 7 Cernan, Evans, Schully

The Future of Manned Mars Missions

Manned Mars missions have been studied as a long-range follow-on to Apollo and Lunar Landings.

Manned Venus landing missions are considered impossible due to the very dense and very hot carbon dioxide atmosphere, the product of a run-away green house gas. Could a run-away green house gas situation occur on Earth? What would a major, nuclear war do to the Earth's atmosphere? International cooperation on scientific missions in space may be a way to prevent major wars.

Manned Jupiter and Saturn missions have also been considered impossible due to the intense trapped radiation zones and cold temperatures, as well as the large distances, leaving Mars as a possible goal. And manned Moon missions are possible during minimal Solar activity, but such missions were considered uninteresting due to the lack of water and a lack of an oxygen atmosphere.

Manned Mars missions are potentially do-able, although more difficult than Apollo. There are several greater difficulties than for Apollo. The thruster requirements for a manned Mars mission has been estimated to be more than three times greater than Apollo, and the mission duration would be longer (one to two years, or more, compared to a few days for a manned Lunar and return mission).

The first big hurdle for a manned Mars mission is the one or two way, 9 to 18 month minimum exposure to the cosmic rays which could be solved by magnetic systems as well as by very thick shielding needed (due to the high energy of the cosmic ray particles in deep space).

Perhaps one should consider making major changes in the Martian atmosphere by artificially warming the polar carbon dioxide to release the frozen water and adding oxygen as a byproduct of the warming process. Such a massive temperature increase for Mars might be created by directing meteor and comet particle impacts onto the polar ice caps. Mars used to be warm enough to have flowing water on the equator. There are several ways that could have happened (natural volcanism and warming of the carbon dioxide, now frozen in the polar caps). How long would that take? Possibly it might be rather quick, less that a normal geological (oops – Areological time scale). Geo' applies to Earth; Ares', to Mars.

To support a long-duration manned Mars mission, and a major habitat to support life, and a transportation system should be built before launching a manned Mars mission. Then the systems for the return to Earth needs to be developed, assuming the astronauts had survived and wished to return.

The discussions of how an astronaut might feel if he had an option of a long-duration, costly flight back to Earth, or to just keep on working until the end, like a battery that will no longer take a charge, that problem is beyond the scope of this book. Were a young scientist and pilot-astronaut to agree to go on a one-way mission to Mars, gets ill, and changes his or her mind and requests a spacecraft to return to Earth, what would happen. Could that request be denied? How long would it take to develop and fund a return spacecraft? As an employee, could an astronaut be denied a return voyage? One way to answer such a hypothetical problem, would be to land a return vehicle and fuel for a return, prior to launch of the initial, manned, Mars landing. Would it not be important to make the Mars atmosphere livable, with water, oxygen, and all other systems for life support, prior to the launch of the Manned Mars mission?

Reference:

Man in Space, by Dr. Piantagosi.

Naomi Oreskes and Erik Conway, the Collapse of Western Civilization, 2014

This Humble Author, painfully disagrees with this kind of Collapse. I am an Optimist. I feel the authors of the above mentioned book, which brilliantly describes the conditions leading to the collapse of the western civilization brought about by a failure to solve the problems of global warming, which this humble author feels that several solutions might prevent such a collapse by as early as 2200.

This humble author believes with some high degree of certainty that several highly predictable things might happen. First and rather important, both the United States of America and China may have exhausted the world's supply of fossil fuels, perhaps as early as 2200 AD. Secondly, human-kind is rather resourceful. I think that there are other sources of energy than fossil fuels. Such energy sources might be very costly, and might result in a drastic reduction in world population. Research on carbon dioxide and methane scrubbers are being developed as a viable method to reduce climate change.

Now that an improbable Mars mission movie has been made, the public's interesting is increasing, but many new unmanned landing missions will be needed before a manned landing on Mars.

Material Loss Due to Fires, and other causes:

In prehistoric times, printing was by hand written books on clay tablets, or on paper scrolls. In modern times, printed book materials in the early stages on computers can still be lost due to erroneous keyboard instructions. It is important to record important texts and images onto what is called thumb drives or flash drives. These drives are purchased separately and are to be plugged into small, rectangular openings, containing electrical connections. Use only one flash drive at a time. Follow the warning instructions for removal of the flash drive, or wait until the computer has been turned off.

MODERN DEVELOPMENTS IN THE LATE 20TH-CENTURY AND THE POST-2000 YEAR TIME PERIOD

Among the most important discoveries after the end of World War II, the Korean War, and the Viet- Nam wars, was the development of the transistor and other semiconductor devices, the silicon solar cell, and the discovery of quarks, Leptons, and other sub-nuclear particles, as well as the discovery of the trapped radiation zones around Earth, Jupiter and Saturn, and the development of the Apollo program and other manned and unmanned space missions, and many medical discoveries and new medical operations. In Astronomy, the Palomar 200-inch telescope was a monumental achievement over difficulty. It was the largest single-primary mirror telescope in the world. It was the product of tireless work of George Ellery Hale, who stated that the Palomar primary mirror could not be duplicated. The future replacement of the Palomar 200-inch mirror would have to be a multi-mirror. A larger single-primary mirror telescope was built in Russia. It had defective optical surfaces.

Larger telescopes exist on several mountains in Hawaii, the Andes, in Southwestern Europe, and elsewhere. Such Large telescopes use multiple primary mirrors.

Reference:

Wright, Helen, The Great Palomar Telescope,

The Solid State Crystal Diode

Several materials were developed as semiconductor diode rectifiers. The old-style battery charger used vacuum tube rectifiers. These vacuum tube diodes were replaced by more efficient and less expensive solid state diodes. Among the first such devices included the selenium crystal rectifier. Selenium crystals were found to convert light into electricity (they became the first solar cells, although, inefficient).

The Transistor 1947

Three scientists at Bell Labs introduced the transistor in 1947. The first transistor consisted of a germanium crystal and two small electrodes, creating an amplifier that served to make a radio, demonstrated by late 1948. The inventors were John Bardeen, William Shockley, and Walter Brattain. These three scientists were awarded the 1956 Nobel Prize in physics.

The transistor revolutionized the field of electronics from vacuum tube electronics to solid state electronics. The first transistor using thin wires touching a germanium crystal had low power capability, but was soon improved by depositing and growing large area junctions replacing the original thin wire elements. Later developments used silicon crystals in place of the germanium crystal. Decades later developments came with growing large arrays of transistors and other components, plugged into a single circuit board.

The Ruby LASER

The first operating LASER used a ruby rod and a flash lamp. It was developed by Dr. Maiman at Hughes Research Lab, and by others at Bell Labs in 1964. LASER is an acronym for Light Amplification by Stimulated Emission of Radiation. A few new LASER types, the gas LASER, the x-ray LASER, and the tuneable dye LASER, were developed. High-powered LASERS were also developed, capable of cutting metals, and welding.

The Microwave MASER

Maser is an acronym for Microwave Amplification by Stimulated Emission of Radiation.

QUANTUM NECHANICS

Richard Feynman, Murray Gell-Mann and others developed the theories of QED, Quantum Electrodynamics and Quantum Chromo**dynamics, partons, and the quarks leading to the new standard model of physics.**

Murray Gell-Mann and George Zweig independently introduce the idea of QUARKS, one of several new sub-nuclear particles, later found with high-energy accelerators at CERN and other locations.

Terminology of Modern Physics

The introduction of modern physics as an extension of the study of the orbits of our planetary system and the long term motions of nearby and distant stars, starting from Newtonian physical concepts, which has evolved into a new and complicated set of physical terms, describing the newly discovered particles making up the center of atoms. The new terms were expanded from interplanetary bodies to the extreme of very small objects in the center of atoms and molecules.

How did Isaac Newton determine the classical laws of physics from a set of experiments? Consider how difficult it must have been to have come up with his theories, and to have proved them. So many things could have transpired to introduce unknown concepts of friction, and elastic behavior of materials under the impacts of stress.

Try your hand at doing such experiments as Newton and others carried out to develop and prove the laws of classical Newtonian physics. Such steps were difficult, just as difficult has it have been in the early 1900's to have developed the ideas of modern physics.

Read about how Newton went about his experiments, and how beginning physics text books suggest the best ways to repeat the steps needed to prove the concepts of classical physics. In the beginning, terms such as energy, force fields, attraction and repulsion, impact and momentum, had to be defined clearly, and measured accurately. An important term is interaction, where complicated motions and interrelations between objects were studied. Try your hand at such considerations, and compare your ideas with those given in textbooks and World Book Encyclopedia. Then try to apply your suggestions with those used by scientists probing atoms and nuclear particles, undergoing unknown particle interactions.

For example, the term deep inelastic scattering has been applied to measurements of the results from early to late, high energy particle impacts and measurements carried out with the earliest alpha particle impact studies by early English and French scientists, and the recent measurements using the high voltage, electrostatic Van de Graaf, cyclotron, and the recent measurements from the high-power CERN accelerator located on the borders between France and Switzerland.

Deep inelastic scattering is the name given to a process used to probe the interiors of the atomic nuclear particles, and the internal nature of sub-atomic particles, such as the proton, neutron, and quarks.

The terns of hadrons (the baryons, such as protons and neutrons), plus electrons, muons and neutrinos. provided the first convincing evidence of the reality of quarks, which up until that point had been considered by many to be a purely mathematical concept, or imaginative idea. The idea that protons, neutrons, electrons and other sub-atomic particles exist, interact and can be studied. Particles that are extremely small can be hard to measure. The use of high energy accelerators to probe the atomic nucleus had to have been carried out, using larger and more costly accelerators.

Glenn Seaborg developed the first high energy accelerator (the cyclotron) at the University of California. A linear accelerator and the travelling wave accelerator were developed at Stanford University.

The idea that the internal structure of protons, neutrons and other sub-atomic particles too tiny to be measured, was developed. Originally atoms were considered immutable. Later internal structure was found for both protons and neutrons, plus other even smaller particles.

It was a relatively new process, first attempted in the 1960s and 1970s. It was an extension of Rutherford scattering with alpha particles on gold foil, and with alpha particles on aluminum foil by Irene and Joliot Curie, during the 1930-1935 time period, to much higher energies of the scattering particle and to much smaller features of the components of the nucleus.

To explain each part of the terminology, 'scattering' refers to the lepton (electron, muon, etc.) being deflected. Measuring the angles of deflection of protons, gives information about the nature of the process. The term 'inelastic' means that the target absorbs some kinetic energy. In fact, at the very high energies of the accelerated particle used, the target is 'shattered', emitting many new particles.

Physicists define subatomic matter into Fermions and Bosons. Fermions have half-integral angular momentum quantum number (the Spin), and Bosons have integral angular momentum quantum number (Spin). Fermions obey the Fermi-Dirac quantum characteristics (called Statistics); Bosons obey the Bose-Einstein quantum characteristics (called Statistics).

Physicists defined the Hadron as any group of particles composed of Quarks, such as the proton and the neutron. The Boson is any group of particles composed of electrons, and other smaller charged particles.

These particles are hadrons and, to oversimplify, somewhat, the process is interpreted as a constituent quark of the target being 'knocked out' of the target hadron and due to quark confinement the quarks are not actually observed but instead produce the observable particles by hadronization. The 'deep' refers to the high energy of the lepton, which gives it a very short wavelength and hence the ability to probe distances that are small compared with the size of the target hadron—so it can probe 'deep inside' the hadron. Also, note that in the perturbative approximation it is a high-energy photon emitted from the lepton and absorbed by the target hadron which transfers energy to one of its constituent quarks. Quarks can't exist free of an atom. They decay too quickly.

The Standard Model of physics, from Murray Gell-Mann and George Zweig in the 1960s, had been successful in uniting much of the previously several concepts in particle physics into one, relatively straightforward, scheme. In essence, there were three types of particles, leptons, quarks, and gauge bosons.

The leptons, which were low-mass particles such as electrons, neutrinos and their antiparticles. They have integer electric charge, but are composed of sub-atomic particles with fractional charges.

The gauge bosons, which were particles that exchange forces. These ranged from the mass-less, easy-to-detect photon (the carrier of the electro-magnetic force) to the exotic (though still mass-less) gluons that carry the strong nuclear force.

The quarks, which were massive particles that carried fractional electric charges. They are the "building blocks" of the hadrons. They are also the only particles to be affected by the strong interaction.

The leptons had been detected since 1897, when J. J. Thomson had shown that electric current is a flow of electrons. Some bosons were being routinely detected, although the W+, W- and Z0 particles of the electroweak force were only categorically seen in the early 1980s, and gluons were only firmly pinned down at DESY in Hamburg at about the same time. Quarks, however, were still elusive.

Drawing on Rutherford's groundbreaking experiments in the early years of the twentieth century, ideas for detecting quarks were formulated. Rutherford had proved that atoms had a small, massive, charged nucleus at their center by firing alpha particles at atoms in gold foil. Most had gone through with little or no deviation, but a few were deflected through large angles or came right back. This suggested that atoms had internal structure, and a lot of empty space.

In order to probe the interiors of baryons, a small, penetrating and easily produced particle needed to be used. Electrons were ideal for the role, as they are abundant and easily accelerated to high energies due to their electric charge. In 1968, at the Stanford Linear Accelerator Center (SLAC), electrons were fired at protons and neutrons in atomic nuclei. Later experiments were conducted with muons and neutrinos, but the same principles as in the alpha penetration experiments had been followed.

The collision absorbs some kinetic energy, and as such it is inelastic. This is a contrast to Rutherford scattering, which is elastic, i.e. no loss of kinetic energy. The electron emerges from the nucleus, and its trajectory and velocity can be measured.

Analysis of the results led to the following conclusions:

The hadrons (protons and neutrons) have internal structure. Quarks appear to be point charges, as electrons appear to be, with the fractional charges suggested by the Standard Model.

The experiments were important because, not only did they confirm the physical reality of quarks but also proved again that the Standard Model was the correct avenue of research for particle physicists to pursue.

References:

Deep inelastic scattering. Oxford University Physics Department, 2003.

E.D. Bloom et al. (1969). "High-Energy Inelastic e–p Scattering at 6° and 10°". Physical Review Letters 23 (16): 930–934. Bibcode:1969PhRvL..23..930B. doi:10.1103/PhysRevLett.23.930.

M. Breidenbach et al. (1969). "Observed Behavior of Highly Inelastic Electron–Proton Scattering".

Murray Gell-Mann, The Quark and the Jaguar, ibid.

Let us try to understand the early ideas behind the Quark. Deep scattering experiments had indicated that the proton in hydrogen, was found to scatter as if it were composed of three sub-nuclear particles. Neutrons had been observed to decay into an electron from the free state. Free hydrogen (made up of protons) is a simple element found in nature or produced by electrolysis of water to be stable (it would not decay as was the case for free neutrons). Heavy hydrogen containing a proton plus a neutron was stable, but hydrogen nucleus with two neutrons (tritium) was unstable. Helium gas which contained two protons and two neutrons was stable, but could be changed into an alpha particle. As long as the material was stable, the neutrons would not decay.

Neutrons were electrically neutral, but when free, they would decay by admitting an electron, and decaying into a proton and an electron.

So let us think of the needed charge of protons and neutrons to behave in such a way. Let us think that if the proton were composed of two smaller particles, each with a charge of one or two thirds of a negative electron, plus a third particle with a single positive one third of positive charge. When combined into a proton we would get a net charge of plus 4-thirds, less 1 third giving a net charge of plus one. Call these two up-quarks. Next, think of another quark with a negative charge. Fractional charges do not exist in the free state without the support of the surrounding nucleus.

In the case of a proton, we would look at two up-quarks, plus one down-quark, giving a net charge of plus one, that is 2 times plus 2/3 charge plus a negative 1/3 charge = 1 positive charge.

In the case of a neutron, we would look at two down-quarks plus one up-quark, giving a net charge of zero, that is, plus 2/3 minus 2 times 1/3 = 0 charge.

Both the proton and neutron had to be augmented with other quarks needed to hold the product together.

The resultant standard model for the proton and neutron is given in the figure at the end of this chapter.

The Standard Model of Physics is not yet complete, it does not resolve the problems of the Einstein theory of relativity, and it does not include any mention of Dark Matter and Dark Energy which make up 95 percent of the universe. Perhaps, a future edition of this summary book might give us some insight of these missing materials. A chart of the Standard Model of Physics is given at the end of this chapter, and in an appendix.

References:

Murray Gell-Mann, The Quark and the Jaguar,

Hawking, Stephen, W., The Theory of Everything, and the Fate of the Universe, 2007

The Impact of DVD publishing of science journals

I had the problem of storing back issues of several scientific publications, due to lack of shelf space, cost, and keeping the many piles of materials well-organized and kept in order. Missed-placed issues some times were a problem. The lack of a good system for retrieval by subject of older material was a constant problem. My collection of several publications became lost. Many such publications in monthly issues were lost in storage. The idea of DVD release of back issues has important interest for me and my future family. The passing-on of my scientific material to my future members of my family and other friends was important. The cost of saving large volumes of un-cataloged material can become overwhelming. Much may have been lost. Were most of all such material saved in DVD format, such material could have been saved and cataloged.

In the near or immediate future, DVD system stored systems are of importance, for me, and others. The cost of a set of DVD of past issues, not necessarily to me or my future family or their friends, is to be considered. The cost of storage and listing of old issues of my Scientific American, Physical Review Letters, or the massive Journals of Geological Sciences, and American Medical Association publications amounts to rather huge piles of paper, most of which may be of great importance to the future, but many issues could become lost forever, in library clean-up operations. The DVD re-publication has a much greater chance of survival and ease of access. Scientific American, for example is now available for a rather low cost, for all back issues, including the 1888 original magazines.

The standard model of physics, simplified:

Reference: The Quark and the Jaguar, by Murray Gell-Mann

The Proton is composed of 2 up Quarks, and 1 down Quark, plus other parts

The Neutron is composed of 1 up Quark, and 2 down Quarks, plus other parts

HISTORY OF SCIENCE FOR THE FUTURE

It is reasonably predicted by many authors, that the future of global warming will proceed to become much warmer on the average, with more severe storms, and other problems such as sea-level rise as a result of the long-lived carbon dioxide, and other greenhouse gasses, already in the Earth's atmosphere. In 2016, the Earth's environment may already be beyond a tipping point. The term, climate change, implies a cyclical basis for global warming, but Global Warming is not cycle, it is a long-term average of the world-wide weather patterns.

Reference: James Hansen, the Storms of My Grandchildren, ibid.

The supply of fossil fuels may run out by 2200, or become so much more costly, that other sources of energy will have to be found, studied and developed. Consider the future costs of water and food might cause a reduction in human population. The costs of military conflict may also be so costly that local and world-wide conflicts might result in a significant decrease in world conflicts, and might usher an end to such conflicts. There may be less food for the armies, and less fuel to move warships, tanks, and military supplies. Carbon dioxide, methane and global warming might begin to decline, as a result, dream on.

That does not indicate that the major economies of the world will not have to take any action against global warming, which likely would occur early enough in time that effective actions may have to be undertaken, and still possible to work. At which point, the cost of energy usage may require changes in how energy is used.

It is possible that the primary fuel supplies (gas, oil and coal) by 2200 might become so expensive, that scientists will be forced to develop new energy sources. It is possible that the only energy available by 2200 may be solar, wind, ocean-wave, and nuclear fission, and fusion power. It is possible that the main source for agricultural water in the drought-stricken southwest would be by solar or wind-powered desalination plants. Massive water pipelines will have to be built to serve areas in severe drought.

Rather than allowing civilizations to collapse, scientists will be required to solve the problems. Even if a new source of cheap energy is discovered, many future children will be needed to become scientists, and scientific studies will have to be started by family teaching of children to become future good scientists, with home-based studies starting before the first grade. To accomplish the best teaching, the women bearing children may need to be trained to be good scientists, as well as good science teachers by training in science before raising kids. The future of science training of young children, both girls and boys, may be our salvation.

Politicians and business leaders, on the Republican side of the aisle, can be expected to vigorously resist changes in their way of life. But as global warming-driven severe weather becomes much more severe, a change might be that people will begin to demand a more-liberal point of view.

Civilizations that deny schooling of girls, especially in the sciences, might begin to fail earlier than mid- century 2100.

Reference: Malala's book entitled, "I Am Malala", the girl who stood up for education, and was shot in the face by a gunman from the Taliban while she was riding a bus in a city in the area of Swat in northwestern Pakistan, October 12, 2012, editor Christine Lamb, Little, Brown, and Co., 2013. The full name of Malala is Malala Yousafzai. The book gives details of her surgery in Birmingham, England, and details of conditions in Pakistan at the time.

Less than about 50 percent of the public believes, as of 2016, that carbon dioxide increases in the upper atmosphere will become a serious problem in the next 50 years. And if not now, what the world should be doing, to avoid serious problems, later. In 2014, extreme weather events seem to have increased. This year (2015-2016) severe tornadoes, high winds, heavy rains, flooding of streets and homes in the southeast, and droughts in the southwest, have occurred. Although individual instances of severe weather events have occurred, no individual instance of bad weather can be directly related to an increase of global warming. Global warming is related to a long-term average of temperature rise, and the increased probability of severe weather related events. The term climate change applies to a cyclical change in atmospheric conditions.

The probability of severe weather-related events, may be cyclical. But the sudden change from loss of ice in the Arctic and Antarctic coastal waters can induce a sudden change in the long-term probability of severe weather-related events. Severe weather events, extreme hot weather, droughts in the southwest, heavy rain storms, and an increase in the strength of hurricanes, tornadoes, residential street and building flooding, loss of land due to sea-level rise, the increased cost of food, due to lack of water for growing crops in the southwest, and flooding in the northeast are a matters of increased probability. Even extreme cold and heavy rain storms in the winter may be caused by thawing of the Arctic Ocean due to global warming, which seems to be a result of reduction in the Arctic ice which normally reflects the sunlight. Loss of the snow and ice reduces the reflection of sunlight, and increases the amount of solar heating in the Arctic, which, in turn, puts more energy into the Polar Vortex, resulting in heavy winter storms in the central and northeastern United States.

Two well known weather oscillations include the West Asian monsoon, and the Southern Pacific Oscillation, which appears to have a more variable cycle, known as the El Nino.

Which weather problems become the more serious depends on the location. For example, beach property on the southeastern United States of America, and low altitude Pacific islands would be affected by sea-lever rise. Property in the central planes would be affected by increased severe tornadoes, and flooding. Coastal areas from Florida to New England would be impacted by increased hurricane intensity. And southwestern US would be affected by increased heat and drought. Agricultural regions in the southwest are being affected by drought, already.

Southern California might seem to be an ideal place to live, excepting the cost of real estate and the fear of the next big earthquake. Each community has its own list of potential problems and solutions, connected to global warming, and other weather-related problems. Scientists have taken the concept that global warming does not cause hurricanes or tornadoes to occur directly, but may cause the average intensity of storms to increase. The Hawaiian Islands may be a very good alternative to Southern California. The Hawaii or the Big-Island of Hawaii are smaller in living area than Southern California, and are lacking in the extensive mineral wealth of California, Utah, and Arizona. Although the Big-Island has extensive active volcanic regions, it may be devoid of the major seismic San Andreas Fault which runs from the Pacific Ocean down past San Francisco to the border of Baja California, Mexico.

Presently, one solution to one type of event, can also create an increased hazard to another type of an event. For example, constructing underground shelters for tornado protection, increases the potential for flooding of the shelters. If many families living in the central to eastern US become tired of flooded roads, high humidity in summer, and shoveling snow in the winter, move to southern California with its warm winters, dry and cool summer evenings, be prepared for higher home prices, earthquakes, and a possibility of over-population.

In the southwest, drought would affect agriculture. To protect our food supply, massive water pipelines and new water delivery systems will have to be built to support the Central Valley agriculture, which is needed not only to feed an increased population in California, but also to provide agricultural products to feed people living in the East and Central states impacted by flooding and extreme storms and higher summer temperatures.

If ocean water levels increase, flooding of coastal communities would lead to migration of people living in regions affected by the ocean level rise. If many families are displaced, new living areas may have to be made in California, Hawaii, Arizona. and elsewhere.

The impact on Mexico, and Central America is beyond the scope of this book. Remember that Central America to South America, and Antarctica have summer when the Eastern and Central have winter.

As coastal flooding begins to impact the southeastern coasts of the US, some families will have to move.

The Polar Vortex

The central northeast regions in the United States have suffered from unusually severe cold, heavy snow, and flooding during the fall and winter in 2014 to 2016. The Polar Vortex is expected to increase starting in April 2016 with high winds and low temperatures, just as spring and warmer weather was expected. This may be temporary, and would seem to reduce the effects of temperature increases. This Polar Vortex will keep the cold temperatures and will delay the arrival of summer weather in the northeast. The temperatures will retard planting of summer vegetables in the central to northeast this spring. This should increase the need for California produce, and the need for greater water supply.

What temperature increases and summer storms can be predicted for summer 2016 remains to be seen at this time. The Polar Vortex is increasing, because global warming is increasing the amount of energy in the Arctic air, as a result of the loss of Arctic ice. The Arctic ice reflects the sun. The loss of Arctic ice causes an increase in energy deposited in the Arctic Ocean. This increased amount of energy in the Arctic causes the spring-time Polar Vortex to increase, bringing unusual cold temperatures to blow into the eastern and central US. Means for reduction of the chance of brush fires, needs study.

A possible result might be to cause the seat of the US Central government to move, temporarily to California, Arizona, or Hawaii, during severe weather, street flooding, in the east coast.

Notes:

The Future Problems of Global Warming

In the 1800's to the 1960's global warming was considered a good thing. Increased carbon dioxide in the atmosphere would be beneficial to agriculture, and a warmer and rainier weather, making for more and stronger agriculture. But increased emissions of fossil carbon can eventually prove hazardous, if it becomes an extreme. Numerous authors have published good scientific books on the effects of future increases. Several well-written books concern the 2100-2200-year range, if no actions are taken to reduce the rate of burning of fossil fuels.

A recent development in the eastern region of the United States is the extreme weather. It might be that the melting of the polar ice has allowed sunlight to warm portions of the Arctic Ocean, once mostly frozen in the winter. The loss of ice which reflects sunlight, allows the sun to warm the Arctic Ocean, leading to an increased amount of energy available to power the weather in the winter. Whether this change is a long- or short-term remains to be seen.

A recently-published book written by Dr. Naomi Oreskes, of the University of California, San Diego, and Dr. Erick Conway of JPL, deals with an imagined collapse of the western civilization after 2200, if nothing is done to reduce the rate of fossil fuel consumption. Another book by Dr. James Hansen, the Storms of My Grandchildren, deals with the possibility of a more recent increase of severe weather.

Hansen, James, The Storms of My Grandchildren, ibid

This humble author's comment is that it seems that such severe storms have begun already, and are predicted to increase, according to excellent computer program projections.

Alternative Energy Sources

At the present time of 2014, there are a few kinds of renewable sources (non-fossil sources of energy) as listed in the following table:

Renewable Energy Sources

Solar cell panels

Solar Thermal

Ocean Wave and Wind Turbines

Nuclear (especially fusion power)

As of 2016, the amount of renewable energy use is rather small compared with gas, oil, and coal. All of the above electrical power sources seem to have hard-to-solve problems. Solar cell power requires batteries for use at night. The wind does not always blow. Ocean wave power can be damaged by sudden waves during hurricanes, rogue waves, and earthquakes. And nuclear power has a problem with depleted fuel-rod storage. And fusion power stations need much further technical development.

Solar cell power panels remain as the most promising source of safe energy, based on future development of high-efficiency cells. High efficiency batteries are being developed. The development of better lighting, such as with LED lamps, looks very promising. The LED uses low voltage, direct current power, which would allow direct connection to solar cell panels without change to alternating current. The LED light panels could be powered by small, low-voltage batteries for use during the night. Residential street lighting with curb-level LED lights could be developed to replace costly sodium-vapor lamps on expensive street light poles.

Solar panels do not become radioactive and can be recycled. The solar cells could be integrated into the roofing materials. Instead of replacing the roof, and adding solar cell panels, use solar cell as the roofing tiles, directly. Solar panels are water-proof. They just have to be kept clean and blown free of leaves.

South-facing building walls and fencing surfaces could be solar-paneled, and never need to be painted. Not having to paint a wall covered with solar cells, is a poor advantage, of course, as the real purpose of the solar cell panels is to produce electricity. The cost of the solar cell panels is coming down due to mass production. How far will the cost drop, and how soon, are beyond the scope of this book.

A more important question is how soon will more efficient solar panels be available, and how soon will the cost become competitive with other sources of energy. At some point in time, would high-efficiency cells become cheap enough to be placed on the roofs of SUV's, busses, and trains.

Presently, in Europe, solar panels are being placed alongside railroads. Charging stations are being added for automotive use. As the world begins to run very low on fossil fuels, costs would rise. Hybrid power for home heating and automotive needs could be met by solar, wind power, and fusion power. But large trucks and busses would have to switch to natural gas, and away from diesel.

As the cost of fuel for overly-large, sport-utility vehicles (SUV's), becomes prohibitive, many families would begin to switch to small, electrical cars to drive a short distance to school or to the market for food and small-volume household supplies. For a few, small items, take a local bus. For larger items, hire a truck. To go to a library or a Doctor's appointment, take a local bus, a small electrical vehicle, or walk (in good weather).

I choose to use public transportation. In Southern California, as in New York, there are busses and trains, fueled by over-head electrical wires, natural gas, or Diesel. The busses and trains run on a schedule, whether or not an individual passenger is on board, hence an individual passenger riding on board does not contribute directly to global warming, excepting for very short trips.

A heavy bus has to stop, and idle while passengers get on or off. Light rail trains, however, have to come to a full stop at each station to let a group of passengers on or off. Hence a several-car, light rail electric train system can be more cost and energy efficient than a typical bus system.

HISTORY OF SCIENCE AND MEDICAL DISCOVERIES

References: Cartwright, Frederick, J., Disease and History, 1972

Dr. Hans Zinsser, Rats, Lice, and History, Bantam Books, 1935

The Plague, Yersinia pestis, begins with flu-like symptoms; high mortality, unless treated with antibiotics.

ANCIENT MEDICINE

Prior to 1800 medicine was mostly a matter of herbal treatment and isolation of the very ill. The average lifetime of humans was about 30 years. Death during childbirth of both mother and child was all too common. Human populations lived near relatively unpolluted streams and rivers. Drinking water was purified by boiling. The safest water for human consumption was rain water. Medicines for common illnesses were very limited. The early understanding of disease as a result of a lack of sanitation and ways to avoid insect carried diseases of malaria and typhus were unavailable, until the mid-1900's.

MEDICAL DISCOVERIES IN THE 1800 TO EARLY 2000 TIME PERIOD

Notable medical doctors (I am not a medical doctor, and will use specialists in future editions of this book).

The early Greek physicians were associated with Hippocrates. The Hippocratic Oath is still in use in medical schools. Advice to doctors at the time of Hippocrates was help your patients, but do no harm.

Edward Jenner (1749-1823) The earliest developer of immunology, with the Small Pox vaccine. The importance of Jenner's small pox vaccine started and revolutionized immunology, and save more children's lives than any other doctor at that time.

Antoine van Leeuwanhoek (1632-1723) Invented the microscope, and imaged cells of the human body, including microbes. He developed techniques to identify diseases by their structure seen by the microscope. He was considered the father of microbiology.

Ignaz Phillip Semmelweiss (1818-1865) stressed hand washing as a requirement between treatment of patients, and that was found to save lives. Hand washing with soap, at that time, was prepared with lye and kitchen fat, and was rather harsh and unpleasant on the skin.

Louis Pasteur (1822-1895) developed the theory that germs were a cause of many diseases, and developed the process of pasteurization for milk, wine, and beer.

Heinrich Robert Koch (1843-1910) Identified the tuberculosis organism.

Joseph Lister (from about 1843-1897) Developed the science of antiseptic surgery near the turn of the century, following the work done by Pasteur.

Alois Alzheimer (1906-1960) proved that Alzheimer's disease was distinct from normal aging. No clear cause and no cure before 2020, when a simple pill to reduce the damage to the brain is expected. There is some idea that a Mediterranean diet might help.

No cure has been found for this disease that requires full time care. My mother suffered from Alzheimer's disease, and died in 1987. I had a problem with the diagnosis. What were the symptoms?

James Parkinson 1817 showed that the shaking of a body in old age was not a normal process of old age, but was a brain disease different than Alzheimer's. My wife suffered and died in 2014 from Parkinson's which required near full-time nursing care. If a family is not wealthy, how can the costs be paid.

Huntington's chorea 1846 (is an inherited disorder)

Development of antiseptic alcohols for cleansing wounds was introduced by about 1900.

Development of hypodermic needles and syringes was developed in the late 1800's.

Sigmund Schlomo Freud (1856-1939)

Diseases studied by numerous physicians: (which are beyond the scope of this book, as each deserves its own book)

Malaria, typhoid fever, yellow fever, pneumonia, cancer, black plague, dengue, meningitis, Multiple sclerosis, Hemophilia, rabies, sexually transmitted diseases, arthritis, flu, mumps, whooping cough, poliomyelitis, peritonitis, pancreatitis, radiation induced diseases, glaucoma, polycystic kidney disease, etc. I have had blood relatives or close friends with Alzheimer's, Parkinson's, Polycystic Kidney disease, diabetes, and Hemophilia. Each of the above deserves its own chapter or an entire book.

Cartwright, Frederick J., Disease and History, 1972

Bhopal, R. S., The Concepts of Epidemiology, Oxford Univ. Press, 2002

Dr. Semmelweiss in about 1840, found that simple hand washing between patients saved many lives in the maternity ward in the Hospital of Vienna. His ideas were largely ignored until Louis Pasteur identified germs and many diseases transmitted between patients and doctors, could be prevented by simple hand washing.

Surgical rubber or latex gloves were developed in the late 1800's by Goodyear Tire and Rubber in Akron Ohio. This important development in a factory that was owned, by Frank A. Seiberling and his two younger brothers, all of whom were friends of my family. A grandson ran for the governor of Ohio during the days before the state came under the control of Republicans. Harvey Firestone was also a friend of my family. My father worked for Firestone and was in charge at the company for pricing and policy. Firestone specialized in the manufacture of high-quality rubber tires, while Seiberling produced other products in addition to tires. Goodyear developed the blimp, a non-rigid helium-filled, lighter than air, craft. Harvey Firestone established in the country of Liberia, devoted to extracting latex from Hevea braziliensis trees, grown from seeds taken from the Amazon area, and grew the trees just in time for World War II. Firestone developed a blow-out-proof tire using a double tube (a tube within a tube) that would allow the driver to reach a repair garage at normal speeds. Those tires were popular on FBI and other police vehicles. Tires with tubes were the norm, prior to the present tubeless steel and fabric belted tires. Tubeless tires for automobiles developed less heat during use, thus would last longer.

Louis Pasteur in the mid-1800's was instrumental in proving that germs and the lack of hand sanitization were responsible for many diseases, and deaths. Pasteur carried on his life-saving work, without a medical degree. The French leaders had to cooperate with Pasteur, along with French politicians. Pasteur developed a method using elevated temperature spraying to eliminate medical problems with wine, beer, and milk, a process of pasteurization, which is now a requirement.

A short list of diseases and several modern medications is given below.

Childhood Diseases including Diseases that Can Affect Adults

Measles, and an adult form of the disease called shingles, which causes severe rashes.

Whooping Cough

Chicken Pox. Can be related to shingles in later life. There is a vaccine that one can get, taken once.

Mumps. Mumps can develop together, or separately on each side of the face. It is painful in adult age.

Poliomyelitis

Hemophilia

Scarlet fever

Adult Diseases that can Affect Children:

Pneumonia, a lung disease

Multiple Sclerosis

Cancers

Polycystic kidney-liver disease (PKD) (an inherited disease)

Heart disease

Tuberculosis

Lung disease

Blood Cancer

Acute and Chronic Leukemia

Breast cancer

Testicular cancer

Pancreatic cancer, a fast-moving cancer which can be fatal within 15 months.

Lung cancer

Heart and Lung Embolism

Brain cancer

Alzheimer's disease: there is much work in developing a medication for this disease. A pill may be available by 2020.

Parkinson's disease. There is no cure at this time.

Lou Gehrig's disease (ALS)

Tropical diseases:

Typhus (spread from Lice on rats and mice)

Malaria, a mosquito-born disease

Ebola fever

Sleeping sickness

Yellow fever

West Nile Virus

Zika virus

Diabetes

Stomach cancer

Arthritis

Osteoarthritis

Bone diseases

Bone Cancer

Asthma

Bronchitis

Sexually transmitted diseases:

Syphilis

Gonnorhea

Genital Herpes

HIV

Oral cold sores

Medicines developed in the 20th century include Sulfa drugs and Penicillin

Penicillin discovery is attributed to Alexander Fleming in 1928

A Nobel prize in physiology was awarded to Sir Alexander Fleming, Sir Howard Walker Florey, and Ernst Boris Chain for work in connection with penicillin in 1945. George Beadle received a Nobel prize in 1958.

Several improvements such as the scanning and tunneling, optical microscopes were developed late the 20th Century.

The Electron Microscope was invented early in the 20th Century, but needed many improvements.

Genetics, the Human Chromosomes, and DNA

The subject of DNA is too complex for a simple presentation. Refer to the many books and articles on the internet for this subject.

Ridley, Matt: Francis Crick, Discoverer of the Genetic Code, Harper Collins Books, 2006

This book is a history in the life of Francis Crick. It gives some technical details of Crick's role in the development of DNA. For technical details, refer to the internet.

An approximate comparison of the diameter or widths of various organic objects, consider the following:

The diameter of human hairs is about 20 micro-inches, or in the following list, 50 micrometers for hairs.

The wavelength of green light is approximately 50 nanometers. The human cell diameter is about 500 nanometers. The diameter of a chromosome is about 300 nanometers. The diameter or width of a human DNA molecule is about 23 nanometers.

Vision Research

Vision research is such an important subject that an entire chapter might be needed in future editions of this history book. There are recent developments in this subject that need to be covered by specialists.

Human Hearing and Vision Research

Studies related to human hearing and vision should be covered, separately, in future chapters of this book. Fifty years ago, medical specialists covered the subjects of eye, ear, nose and throat. Brain studies and mental research (brain diseases and psychology) were in separate fields. More recently, hearing specialists tell us that humans hear and see partly with the brain. About 5 years ago, I had an appointment with a doctor specializing in allergy problems. That doctor gave me a hearing examination, as that procedure was his routine. He was a specialist in the areas of eye, ear, nose and throat.

I knew a medical specialist, renting a home on San Marino Avenue in Pasadena, California. I may ask this person, and or others with medical degrees to write more for this chapter, in the next edition of this book.

The plague is still around. About 200 cases per year, worldwide, have been identified. One must avoid touching dead rodents in the wild that may still have fleas, that may carry the plague to humans. Early symptoms are similar to the common influenza. Treatment with antibiotics, is successful, if started early.

History of Science: the Bubonic Plagues, other major pandemics, and the Crusades (dates are approximate and unchecked)

The bubonic plague, also known as the black death, or specifically Versinia pestis are listed below.

The plague of Justinian, 6 and 7th century AD, approximately 570 AD

40 percent deaths in Constantinoble, and in Europe in 1603

Third Pandemia in China in 1890

Italian Plague in 1629 to 1681

Seville plague in 1647 and 1657

The Plague in London 1665 to 1666

The Plague in Vienna 1679

The plague in Magrinoble in 1720 to 1728

The plague in eastern Europe in 1738

The plague in Russia in 1770 to 1772

The Cholera in the 19th and 20th century

First Crusade in 1095 to 1099

Second 1147 to 1149

Third 1187 to 1192

Fourth 1202 to 1204

Fifth 1217 to 1221

Sixth 1228 to 1229

Seventh 1248 to 1254

Eighth 1269 to 1270

Ninth 1271 to 1272

References: The Black Death

Cartwright, F., Disease and History

World Book Encyclopedia

APPENDIX U

Reference: the Chemical Rubber Handbook

IMPORTANT DEFINITIONS OF PRIMARY UNITS OF MEASUREMENTS

Scientific work generally requires the metric system. Engineering and architecture use the English system of units in the United States. Only the US and two other countries use the English system.

The metric system in the US is used in aircraft design and in the automotive industry. Bolts, and nuts are available in stores such as Home Depot primarily in the English system, but some hardware items are also in metric units. Use of the English system of units has caused problems when engines and tools are purchased from Europe, Japan and China and most other countries. Along with problems with dimensions, tolerances may be troublesome. The insistence of using engine thrust in US pounds has caused difficulties. NASA requires use of the metric system by all suppliers, some of which have resisted, or have ignored the requirement. Expensive spacecraft have crashed as a result from errors.

US engineers have a good 'feel' for weights in pounds and for dimensions of parts in inches. Distances measured in inches or centimeters can be understood simply by the relation of 2.54 centimeters equals one inch, exactly.

An example of a calculation of speed in English and the metric system is as follows:

60 miles/hour x 2.54 cm/inch x 12 inch/foot x 5280 feet/mile / 100 cm/meter / 1000 meter/km)=96.56064 kilometer/hour

Note that one can check the calculation by treating the unit names as an equation as follows.

Miles/hour x 5280 feet/mile x 12 inch/foot x 2.54 cm/inch / (100 cm/meter x 1000 meters/kilometer)=

1.609344 kilometers/mile

When renting a car outside of the US, request a card showing the relation of mph to km/hr and collect such a card as a souvenir, in several languages for each country visited.

The kinetic energy relation is 0.5 times the mass times velocity squared, in the metric system. In the English system kinetic energy is 0.5 x (pound/32.174 cm/sec squared) x the velocity squared, and is for measurements on the surface of the Earth where 32.174 is the acceleration of gravity. But when calculations for a body falling from a great height, where the acceleration of gravity varies, and resistance of the air flowing around the body, one encounters some rather complex equations. Learn the proper way to type scientific equations, using clear scientific notation including the unit names.

An interesting experiment on the speed of a falling body is as follows. The standard lab experiment consists of a paper spark strip and a small ball dropped from a height of about 1 meter. A more complex experiment is to use a digital motion picture camera to photograph a series of balls dropped from a multi-level parking garage roof. Compare the result for a white painted marble-sized steel ball, a round block of wood, and a ping pong ball. Do it at night when wind is zero, and in the day. Check the formula with the standard s = g times t-squared/2. Compute the effects of air resistance, if important. Carry out the experiment safely. Appoint inspectors on the roof and on the ground, to prevent accidents Avoid leaning over the railing to look from above. A ball drops about 16 feet in the first second, and 64 feet the next second. The velocity after falling for two seconds is given by $v^2 = 2 \times 32.174 \times 64 = 64.174$ feet/sec.

Human survival after falling more than a second onto a hard surface is near zero.

A safer and easier experiment is to time a round or cylindrical object rolling down a slanted board at various angles from the vertical. Compare results with the object in free fall.

Units in the United States of America:

2.205 or 2.20462 pounds per kilogram, approximately, and at sea level.

The pound is a unit of force, of gravity, at sea level. The kilogram is a unit of mass.

1 inch = 2.54 centimeters, exactly

1 mile = 5280 feet

1 jigger is 1.5 ounces, liquid US measure

1 cup is 8 ounces, liquid US measure

4 cups is 1 quart, liquid US measure

4 quarts is 1 gallon, liquid US measure

1 gallon, liquid US measure, measures 3.78541 liters

a 2x4 (in inches, approx.) of Douglas fir, 10 feet long, is a US standard item (stud) of structural wood

The scientific name of Douglas Fir is Pseutotsuga menziesii

0 degrees Kelvin = - 273.15 degrees Celsius

T degrees Celsius = $5 \times (F-32)/9$ for F degrees Fahrenheit

Speed of light in a vacuum $c = 299792468$ meter/second

The acceleration of gravity on Earth, on the equator, at sea level is 32.174 feet per second-squared

In the American system of units:

1 pound or 1 pound Avoirdupois is a force unit valid at sea level near the equator of the Earth.

1 slug is the American unit of mass. There are various units of mass and force available in the US.

Use of the pound for a mass unit leads to trouble, causing lots of mistakes. The pound is a unit of force.

The pound should never be used as a unit for mass.

Use of the non-scientific ratio of pounds per kilogram, 2.205 pounds, approximately, per kilogram is a good, though a mixed-up unit. It seems to be an 'apple-orange' unit of a force and a mass ratio.

The Table of Units in the Chemical Rubber handbook is Copyrighted. All units given in such tables are Copyrighted. It is considered that you must request permission to copy copyrighted material. If you do use such copyrighted units, in a table, which author needs to be contacted for permission? I feel that such tables of units should be printed in a book with a statement that the table of units is not included in the copyright for that book.

Let us dream that there exists a government agency that assumes the role of serving the public for such a table, for an example, let us dream that the US Supreme Court should take on such a role, and given out the rights free to the public, somehow, and take on the responsibility of providing the best available set of current units, within a given time for a renewal, say, yearly. The same idea should be followed by other such tables as the Periodic Table of the Elements, most of which should be free for public use. Let at future Supreme Court take over the responsibility and publish their best available set of tables.

There are several copyrighted class of books for which a copyrighted table of units may be ignored, those prepared by an author over a 100 years ago, after the author's, death, and published long after the author's life, and printed by a publisher no longer in business.

An example of a Table (of all the known elements), presently available on the internet, is the following Reference: a Dynamic Periodical Table of Elements, by Michael Dayah, on the internet. This appears to be one of the better (interactive) periodic tables available, but one should not try to copy it for publication, except by the author's written permission.

Isaac Newton's equations and constants:

Gravitational force between two bodies in space $F = Gm_1m_2/(r\text{-squared})$, where $G = 6.70 \times 10$ to the neg. 8

Gravitational acceleration = 32.174 ft/sec-squared and 980.665 cm/sec-squared at sea level

1 Acre = 43560 feet there are 640 acres in a square mile

0.01639 liters = 1 cubic inch and 1 liter = 61.0128 cubic inches

www.ingramcontent.com/pod-product-compliance
Lightning Source LLC
Chambersburg PA
CBHW030847180526
45163CB00004B/1483